The Cooling

The

LOWELL PONTE

PRENTICE-HALL, INC.

Cooling

Englewood Cliffs, N.J.

Figure 4-2 and Table 7-1 from *The Weather Machine* by Nigel Calder. Copyright © 1974 by Nigel Calder. Reprinted by permission of The Viking Press, Inc.

Figure 3-3 from Vahe Kirishjian for *Fortune Magazine*. Reprinted by permission; © 1974 Time Inc.

Figure 4-1 from "Ancient Temperatures" by Cesare Emiliani in *Scientific American*, February 1958. Reprinted by permission of Scientific American, Inc.

Figure 13-1 from "Climate Stabilization: For Better or for Worse?" by W. W. Kellogg and S. H. Schneider, from *Science*, Vol. 186, pp. 1163–72, December 27, 1974. Copyright © 1974 by the American Association for the Advancement of Science. All rights reserved. Used by permission of AAAS and authors.

Printed in the United States of America

Prentice-Hall International, Inc., London
Prentice-Hall of Australia, Pty. Ltd., Sydney
Prentice-Hall of Canada, Ltd., Toronto
Prentice-Hall of India Private Ltd., New Delhi
Prentice-Hall of Japan, Inc., Tokyo

10 9 8 7 6 5 4 3 2 1

Library of Congress Cataloging in Publication Data
Ponte, Lowell
   The cooling.

   Bibliography: p.
   Includes index.
   1. Climatic changes.   2. Cooling.   3. Weather control.   I. Title.
QC981.8.C5P66   551.6   76-7963
ISBN 0-13-172312-X

Contents

Acknowledgments

Many scientists and other specialists provided direct and indirect help in the writing of this book. I regret that a short list cannot include them all. The following deserve special appreciation: Lester R. Brown, Dr. Reid A. Bryson, Dr. Pierce S. Corden, Dr. Cesare Emiliani, Dr. Norman Fertig, Dr. William A. Fowler, Dr. W. Lawrence Gates and his co-workers at the RAND Corporation, Philip J. Klass, Dr. George J. Kukla, Dr. Robert Leachman, Senator Claiborne Pell and his staff members David McKillop, Richard Haass and Geryld Christianson, Congressman Gilbert Gude and his assistant William Reinsch, Congressman Donald M. Fraser, his assistant Ruth Knight, and Richard Mauzy of his subcommittee staff, Dr. Walter Orr Roberts, Dr. Stephen H. Schneider, Dr. William Van Cleave, and all those in the defense community who asked not to be named.

Belated thanks to Dr. Clyde L. Cowan (1919-1974) for the years of fruitful correspondence, and to his widow Betty Cowan for sharing her husband's documents. He was full of passion, compassion, honesty, and curiosity, a rare and precious combination. Without his stimulus this book would never have been written; hence it is dedicated to him.

Special thanks to Mrs. Betty Cowan for permission to quote from a speech prepared by her late husband, Dr. Clyde L. Cowan, in conjunction with Richard Nixon's staff in 1968.

Likewise this book would have been far poorer without the ceaseless labors of the public librarians of Santa Monica and Beverly Hills, who I think must midwife as many books as they shelve. Thank you.

And thanks to the media-amigos and other friends who kept me writing and provided helpful comments and ideas. In the order of the alphabet: Christopher Ames, Maxine Asher, Angela Masson Barris, Peter Bunzel, Hank Burnett, Jr., Larry Burns, Ron Cobb and Robin Love, Randy Darden, Jim Doyle, Barry Farber, Chris Fecht, Steve Futterman, Barry Gray, Ken Grubbs, Carol Hemingway, Taylor Hackford, Chuck and Mary House, Kendall Johnson, Bob Kephart, Bonnie and Jonathan List, Lynne Littman, John Lowry, The McNaught Syndicate Gang, Richard Ney, Eric Protter, Alan and Wendy Riche, Dana Rohrabacher, Chris Schlumberger, Del Schrader, Art Seidenbaum, Ted and Wina Sturgeon, Bob Trounson, Dennis Turner and Eva Olenmark, Nick Von Hoffman, James Vowell, and John Ware.

Without my wife, Ellen Levenson, all that follows would have been impossible. Without my parents, and their love, where would I be now? It is better, all three have shown me, to spread a little light than to cast a large shadow. While it would have been easy to write a book on this topic that spreads heat, I have striven to spread light instead.

All errors and shortcomings herein are solely the responsibility of the author.

Foreword

BY SENATOR CLAIBORNE PELL
CHAIRMAN, SUBCOMMITTEE ON OCEANS
AND INTERNATIONAL ENVIRONMENT

The enormous power and capricious behavior of the Earth's environment have amazed and terrified man throughout the ages. In various ways man has searched for methods to control the vast climatological and geophysical forces which have awed and ravaged him. Primitive shamans used incantations and talismans; modern scientists seed clouds and experiment with ways to produce geothermal energy.

Man's attempts to master and manipulate his environment have not always been for peaceful purposes, but it is only in the last several years that we have come to realize the potential horror involved in harnessing natural forces for hostile purposes.

Mr. Ponte's book is a fascinating and important contribution to the growing literature of what has come to be known as environmental modification, or Enmod in the acronymic vocabulary of the arms control bureaucracy. What distinguishes Mr. Ponte's work is his thesis, which is bound to be controversial, that changes in the environment—in this case the natural cooling of our planet's climate since 1945—can constitute a source as well as a means of conflict among nations.

If, indeed, the climatological changes which Mr. Ponte foresees do in fact take place—that is if the cooling produces bad weather and widespread crop failures—then the world's leaders must come to grips with the real possibility, as Mr. Ponte contends, that food will very soon play a dominant role in world politics and that many of a cooling world's nutritionally disadvantaged nations will seek to develop nuclear or other weapons of mass destruction, including environmental modification weapons, in order to increase their bargaining power in the struggle against famine.

Given our own government's recent saber rattling reaction to the prospect of another oil embargo, Mr. Ponte's warning is not as far-fetched as it may seem. Regardless of whether one finds the specific scenarios he develops to be realistic, Mr. Ponte has demonstrated that the provocative issue of environmental politics deserves further informed public debate and governmental attention.

In setting forth options to deal with the cooling phenomenon, Mr. Ponte sees possible salvation through further research and experimentation, which may—if given priority attention by governments—suggest safe ways in which weather or climate can be changed for the benefit of all mankind. He warns, however, that the nations of the world may devote more energy to developing the destructive rather than the constructive aspects of environmental modification particularly as the realities of the new "cold war" become more manifest. Thus this book is as disquieting as *Silent Spring* in its analysis of environmental hazards that can affect our future. If Mr. Ponte's worst fears come to pass, *The Cooling* could prove to be the most important and prophetic popular science book of the 1970s.

Even without such a cataclysmic stimulus, I find it troubling that environmental modification techniques have already been applied to warfare and that developments within the environmental sciences, particularly in the field of weather modification, are rapidly narrowing the gap between fact and fiction. Deeply troubled by the implications of the Defense Department's weather modification activities in Southeast Asia, I urged as early as 1971 that the United States take the initiative in developing an international agreement banning all forms of environmental warfare. At the same time, I called for a more active role on the part of the United States in international cooperation for the peaceful uses of environmental modification. In my view, the military use of any environmental modification technique can only lead to the development of vastly more dangerous techniques whose unpredictable consequences may cause widespread and irreparable damage to the global environment.

Mr. Ponte has recounted my efforts in behalf of the development of the draft treaty banning environmental modification as a weapon of war which the United States and the Soviet Union tabled in August 1975 at the Geneva Disarmament Conference. If concluded and universally accepted, such a treaty would ensure a peaceful framework for the vital research into the future applications of environmental modification for which Mr. Ponte so persuasively argues.

Preface

BY PROFESSOR REID A. BRYSON
CLIMATOLOGIST AND DIRECTOR OF THE INSTITUTE
FOR ENVIRONMENTAL STUDIES, UNIVERSITY OF
WISCONSIN—MADISON

The Cooling will be controversial, because among scientists most of the matters it deals with are hotly debated. There is no agreement on whether the earth *is* cooling. There is not unanimous agreement on whether it *has* cooled, or one hemisphere has cooled and the other warmed. One would think that there might be consensus about what data there is—but there is not. There is no agreement on the causes of climatic change, or even *why* it should not change among those who so maintain. There is certainly no agreement about what the climate will do in the next century, though there is a majority opinion that it *will* change, more or less, one way or the other. Of that majority, a majority believe that the longer trend will be downward. Nevertheless, it is an important question, as this book points out, and it is time for some of the questions to be settled. Lowell Ponte has summarized the data and theories very well, and has reasonably concluded that a rapid change in Earth's climate is possible, perhaps even likely, within the next few decades, and that this would have serious consequences for mankind.

There is surprisingly little argument among those who have actually studied climates over multi-millennial time scales that we will be in an Ice Age 10,000 years from now. There is, however, less agreement about how soon and how rapidly the transition from the present interglacial will take place. One extreme view envisions a "snow blitz" beginning of the ice-age climate, only a few years long, and a rapid growth of continental glaciers. If this were true, response would be almost impossible. The other extreme is the opinion that climates change gradually and almost imperceptibly over many thousands of years, with plenty of time for adaptation by the ecosystems and man. My own opinion is intermediate—that climates change by relatively abrupt small steps; that these small steps are important, for they can be disruptive to stressed ecosystems such as ours is now; and that man can prepare somewhat for their occurrence.

There is also a great deal of argument about the efficacy of various options in preparing for, or dealing with, a changing climate. After several decades there are still widely divergent opinions on the magnitude of deliberate weather modification effects, especially as practiced or not practiced by the military.

There is no consensus about whether there is still time to let normal agricultural research develop crops and technologies that will "save the world from hunger." There is almost no question of the quality of the research, but a great deal of question as to whether it will be "too little and too late." Nevertheless, the problem is sufficiently important that all promising leads must be followed, and a number of them are outlined in this book.

Not everyone is enamored by the concept of caring for 10, or 20, or 30 billion people in a highly regulated technological society, even if it were possible. I'm not, and I'm not even slightly convinced that it is possible *or* desirable. I am convinced that there is very little in the way of human ill, ecosystem degradation, resource shortage, social stress, and international instability that would not be relieved markedly by having fewer people on earth than there are currently.

But rather than fewer people on Earth, we will have more. With population saturation, any climatic variation becomes important; any option worth considering. This book raises some of the questions, and gives us some of the possible ways that mankind might respond short of eliminating people.

When I first read the manuscript I started to accumulate large numbers of marginal notes. There are very few pages that, as a scientist, I could accept without questions of accuracy, of precision, or of balance. As I read on, I threw away my critical notes and started to record the points the author brought up that I had missed in my reading. Lowell Ponte's overall representation of the problem presents a reasonable picture of the hazardous possibilities we would face if Earth's climate changes significantly. Mr. Ponte has delineated the outline of the tangled jungle which mankind must chart to find a way to the other side. He has put the map of climatic arguments into a reasonable perspective. He has shown that there are potential solutions. I hope that scientists will read it as a challenge to set their theory and analysis in order. I hope that all will read it as a serious and thoughtful analysis of a real and pressing problem.

Breaking the Ice

I live where the land ends, on a beach in what the tourist guides call "sunny Southern California."

The Spanish found it otherwise. When they landed here in 1543 they named this place "the Bay of Smokes" because it seemed always enshrouded in fog; these explorers built their pueblo of Los Angeles a dozen miles inland to escape the dense mists.

Four centuries ago the world was suffering what weather scientists today call "the Little Ice Age," a prolonged period of cold that killed off Viking colonists in Greenland, turned many Northern European villages into ghost towns, and on several occasions froze the River Thames in London.

But by the twentieth century climate was warming. Movie studios flocked to the beach cities of Los Angeles, where neither fog nor rain hampered picture-making. Memories of a colder world were forgotten. Everywhere food production increased, and world population grew apace. Nobody paused to think that Earth's climate might cool again, or ponder what would happen if it did.

I have lived on the beach only a few years, but I soon may move. The fog comes frequently now. Neighbors who have lived many decades here all tell the same story: during the 1940s the fog suddenly got dense here, then slackened; since 1950 each year the fog has come more frequently and has become a bit denser than the year before. "It's almost as if we were becoming San Francisco," jokes one gray-haired friend.

It is no joke, but a literal truth. Since the 1940s the northern half of our planet has been cooling rapidly. Already the effect in the United States is the same as if every city had been picked up by giant hands and set down more than 100 miles closer to the North Pole. If the cooling continues, warned the National Academy of Scientists in 1975, we could possibly witness the beginning of the next Great Ice Age. Conceivably, some of us might live to see huge snowfields remaining year-round in northern regions of the United States and Europe. Probably, we would see mass global famine in our lifetimes, perhaps even within a decade. Since 1970, half a million human beings in northern Africa and Asia have starved because of floods and droughts caused by the cooling climate.

The fog over Southern California is one symptom of a changing world climate. Consider some others:

● In tropical Brazil frost and snow have repeatedly devastated coffee plantations, in 1975 destroying more than seventy percent of the crop, killing many plants, and hence driving up world coffee prices.

● In Africa the Sahara Desert has been marching south at up to 30 miles per year into the Sahel nations. There and in Ethiopia the worst drought in modern memory has caused the starvation of perhaps 400,000 people since 1970. Similar droughts, caused by a worldwide shift in rain patterns, have also occurred in India, Central America, the Caribbean, Venezuela, Northern Australia, and South Africa. Wild swings between flood and drought have hit Bangladesh, Japan, China, and the Mediterranean.

● In Europe the 1960s were years of unusual cold, but the 1970s have featured odd warmth. In 1973 England suffered the driest year since 1749, and in 1975 France had the least rain in seventy years. In summer 1975 Swedish factories shut down as temperatures for a week hovered around 95°F. Like chills and fever, these all are signs of a planet catching climatic cold, warns British climatologist Hubert H. Lamb. In January 1976 western Europe was raked by freakish storms and 100 mile per hour winds.

● In the Soviet Union cold winters have combined with drought and heat waves to cause huge crop failures and "the worst weather in a century," say Russian scientists.

● In Alaska the summer of 1975 saw a fleet of ships carrying supplies for oil pipeline construction in the Arctic trapped in what *Time* Magazine called "the worst ice conditions in 77 years." At the other end of the pipeline, scientists warned, changes in the nearby glacier Columbia could at any moment clog the "ice free" oil-tanker port at Valdez with icebergs.

● In the continental United States severe floods have destroyed billions of dollars' worth of property in the Mississippi Basin, the Great Lakes region, Pennsylvania, and New Jersey, while drought has ruined harvests in the Southwest. The East Coast, like Europe, shivered during the 1960s but has enjoyed unusual warmth since 1970. California was balmy during the 1960s, and has been cooler since then; in 1975 Los Angeles experienced the coolest spring since the 1880s. Following a warm autumn, the 1975-76 winter brought the most severe blizzard in a decade, the earliest in sixteen years, in the north central U.S. and many parts of the nation were hit with what the National Weather Service called "record cold." Rain and flood described as "the worst in a century" struck large areas of Washington State. One hundred mile per hour winds lashed Boulder, Colorado, and snow and frost returned to Florida. In February 1976 San Francisco had the heaviest snowfall in 89 years; California rejoiced, for this storm ended the state's most severe drought in a century.

● In the Northern Hemisphere, the half of our planet north of the equator in which most humans live, glaciers that had been shrinking since the 1880s started to grow after 1950. The rapid advance of some glaciers has threatened human settlements in Alaska, Iceland, Canada, China, and the Soviet Union. Sea ice has again become a threat to ships in the North Atlantic.

During the winter of 1971-72 average annual snow cover suddenly increased by twelve percent and has remained high since then. Seven such winters in a row, warned two leading American weather scientists, and we would have an Ice Age.

Why has our planet been cooling? Scientists argue about this among themselves, and without a scientific consensus political leaders are reluctant to act. But the danger is apparent. The cooling has already killed hundreds of thousands of people in poor nations. It has already made food and fuel more precious, thus increasing the price of everything we buy. If it continues, and no strong measures are taken to deal with it, the cooling will cause world famine, world chaos, and probably world war, and this could all come by the year 2000.

Plans for action already exist—ways to help us feed and power the world despite the cooling, and ways to change weather and climate. I helped develop some of them while working as a Defense Department consultant during the 1960s. Already the Soviet Union has set off at least three nuclear explosions in a program they hope will change the cooling climate. Already the United States has shared research with the Soviets in preparation for a joint project that, if necessary, will be used to fend off a new Ice Age. But such environmental tampering is potentially as dangerous to humankind as the cooling itself. Because we each have a stake in such experiments—because the survival of all human beings depends on them—we deserve to know the facts about Earth's cooling climate and the things that can be done about it.

Outside my window the gray fog washes through an elm tree's branches. Faces seem to appear in it. Did our cave-dweller ancestors see such faces swirling in the blizzards of the last Great Ice Age 20,000 years ago? As if by instinct something stirs inside many of us, warning that a great invisible natural wheel has begun turning. Are we right to fear the cooling?

Between 1880 and 1940 a small tropical animal invaded the United States and migrated as far north as Nebraska. The armadillo, a warmth-loving relative of the opossum, spread his range during this, the warmest climatic period in 800 years. Since 1940, as if guided by an unknown natural sense, this animal has been migrating southward again toward the equator. In Europe a warmth-loving snail species became extinct during the 1960s. In southern Canada several species of plants have begun a southward retreat in their growing range. Cod and other fish in the North Atlantic and North Pacific have headed south into warmer waters. Do they sense a change in the planet we have yet to recognize?

I started writing this book for intuitive, emotional reasons, a profound feeling that our planet was changing and that humankind would be forced to change with it or perish. Perhaps a similar intuition has led you to read thus far. But beyond the feelings that tell us "something is happening" lie curiosity, which makes us ask what exactly is going on, and rationality, which seeks to find out why Earth's climate is changing and what we can do to survive in the face of it. From this page forward this book leaves emotion behind, for cool logic must prevail if we are to understand and cope with an increasingly unstable climate in a politically unstable world.

In the following pages we will explore the evidence that leads many of the world's leading weather scientists to say the chances are one in ten that a Great Ice Age will be upon us in less than one hundred years. We will investigate the reasons world leaders fear nuclear war or environmental warfare if the climatic instability, of which the cooling is a part, continues. And we will take a careful look at the plans they are seriously considering to give humankind technological control over weather and climate. What I discovered while writing this book, and what you will find while reading it, is more dramatic than any fantasy of the emotions. It is cold fact: *the global cooling presents humankind with the most important social, political, and adaptive challenge we have had to deal with for ten thousand years.* Your stake in the decisions we make concerning it is of ultimate importance: the survival of ourselves, our children, our species.

Wherever on Earth you live, the weather has been strange in recent years. Odds are it will grow stranger and stranger if the cooling continues, and against those odds are wagered the human future on this planet. You deserve a voice in how we play our hand against the cooling, and this book can help you find it.

LOWELL PONTE
Santa Monica, California
March 1976

PART I

Forces
That
Change
Climate

1

The Cooling

Our planet's climate has been cooling for the past three decades.

Most experts agree on this, for it has been carefully measured by scattered monitoring stations throughout the world. Climate in the southern half of our planet has been warming rapidly, according to the few measurements available. But in the half of our world north of the equator, where most human beings live, the annual mean atmospheric temperature has plunged by seven-tenths of a degree Celsius,* more than enough to offset the southern warming and to lower the average temperature of the whole planet by one-half degree C. At 65 degrees North Latitude in Iceland, the temperature has dropped by 2°C. Some monitoring stations inside the Arctic Circle report that temperature has been falling by more than 2°C. *per decade* for the past thirty years.

Such numbers are bigger than they seem. At the greatest advance of ice during the last Great Ice Age, 23,000 years ago, when large glaciers appeared as far south as Mexico City, Earth's annual mean temperature was perhaps only 4°C. colder than it is today; in some parts of the Northern Hemisphere it was only 3°C. colder. This mean temperature marks the balance point between the forces of heat and cold on our planet. When it begins to fall, the balance is tipping in favor of cold. A tiny change in mean temperature can have large consequences.

At the start of the last Great Ice Age the climate was much like ours today—a pleasant, warm world. Then came a cooling, too small at first to be noticed by the planet's inhabitants. An initial mean temperature drop of only 2°C. might have been enough to

*1° Celsius = 1.8° Fahrenheit. Celsius is also called Centigrade because it marks 100 steps between the freezing and boiling points of water.

initiate an advance of ice and snow toward the equator. Once in motion, the ice and snow caused more cooling, which created yet more ice and snow. The Ice Age gained momentum. According to Soviet climatologist Mikhail Budyko, if world temperature fell 2°C. again, the next Great Ice Age could begin.

Are we at the dawn of a new Ice Age? In 1975 the U.S. National Academy of Sciences issued a report saying that if the present cooling trend continues there is a "finite" chance an Ice Age could begin "within 100 years." How much chance? The NAS panel, comprised of many diverse expert judgments, could unanimously agree only to set the odds of this happening at no better than one in 10,000. The number was not random. As their report noted, Earth's climate in the past has tended to change in fairly regular cycles, and if the past patterns continue we should now be entering a 10,000 year period of cooling climate.

The NAS report was shocking, for it represented a warning by some of the world's most conservative, prestigious, cautious scientists that an Ice Age beginning in the near future—perhaps even emerging from the cooling trend begun during the 1940s—was not impossible. More than that, the tone of the report was one of repressed alarm. When compared to climate of the past million years or so, said the report, this century's warm climate is "highly abnormal." Moreover, "as we approach the full utilization of the water, land, and air, which supply our food and receive our wastes, we are becoming increasingly dependent on the stability of the present . . . [abnormally warm] climate." The NAS Committee on Climatic Variation urged an immediate near-quadrupling of funds for climatic research. Why? Because, said the NAS report, "We simply cannot afford to be unprepared for either a natural or man-made climatic catastrophe."

Why all this concern for something the NAS oddsmakers say is so unlikely? The answer is that many scientists take the possibility of an Ice Age beginning within the next 100 years very seriously. Scientists interviewed by British science writer Nigel Calder in 1974 said they set the odds against its happening at only ten to one, what climatologist Stephen H. Schneider of the National Center for Atmospheric Research (NCAR) calls "Russian roulette odds."

Other scientists disagree, and some believe we are on the brink of a warming trend in global climate. They could prove right, but we can take cold comfort in that. If this happens it will mean our climate has swung wildly from severe warming during the Dust Bowl era of the 1930s to severe cooling during the 1960s (when Europe and North America experienced some of the coldest winters in living memory) to a sharp re-warming that could begin at any time. This would almost certainly mean that Earth's climate has entered a period of extreme instability.

The cooling is a fact, to be taken as you wish. Is it the forerunner of a new Ice Age? Or is it merely a sign of an unstable, fast-changing climate? This book explores both possibilities, although its bias is toward ice. The point to keep in mind is that the consequences of both may be similar: our death by fire. Either a cooling or wavering climate will hamper world food production as weather gets progressively worse and tends toward extremes: heat waves and cold snaps, floods and droughts, frost and snow in the tropics and bizarre hot weather as far north as Scandinavia. The damage this can cause is already apparent in global food shortages and in the recent deaths of more than 400,000 people in Africa and Asia who starved because the monsoon rains on which they depended have suddenly become undependable. If global famine arises, or if unstable climate threatens the security of any of the growing number of nuclear weapon nations, we can expect world war. Thus, the cooling and all it represents is more than a threat of glaciers visiting our great-great-grandchildren. It threatens all of us right now.

Why has it been happening? Will it continue? Many scientists are working on answers to such questions. Some tentative answers have already been proposed.

Disappearing Daylight

Almost all of Earth's warmth comes from the sun's light. How much would that light need to diminish to plunge Earth into an Ice Age? Not much, Budyko concluded from his computer studies at the Soviet Hydrometeorological Service; a 1.6 percent decrease in the solar energy we receive would be enough to chill the world and start glaciers moving toward the equator. Since 1950, many glaciers have been advancing and growing, and sunlight has been diminishing.

In March 1975 two scientists with the National Oceanic and Atmospheric Administration (NOAA) released a study they had done of sunlight in the contiguous United States between 1950 and 1972. They discovered the duration of direct sunlight had decreased by a net 5 percent, the biggest loss coming after 1964. This, they said, was equivalent to the disappearance of 10 minutes of sunshine a day.

The amount of sunlight a place gets largely determines what its climate is like. How much sunlight it gets depends on how close the place is to Earth's equator, where winter never comes. Nearer Earth's poles the days grow longer in summer, shorter in winter. This is because Earth tilts like a top, spinning on its North–South axis; during its annual trip around the sun the South Pole points toward the sun part of the year (winter in the north) and six months later the North Pole points sunward (summer in the north).

More sunlight enters the atmosphere over the equator than anyplace else on the planet. To find out how other places compare, we can use what University of Wisconsin geographer Glenn Trewartha called "thermal days," using the equator's sunshine as a yardstick. In Quito, Ecuador, on the equator, a year is 365.2 "thermal days" long.

Move north or south and the year shortens. Ten degrees latitude to the north in Caracas, Venezuela, and Addis Ababa, Ethiopia, and to the south in Lima, Peru, the year is only 360.2 thermal days long; five days' worth of solar energy have been lost. Twenty degrees away, near Mecca and Mexico City and Bombay, the year shrinks to 345.2 days. Thirty degrees away, near Cairo, New Delhi, New Orleans, and Durban, South Africa, the year is 321 thermal days long.

The 35-degree latitude lines are energy watersheds on the planet. At lower latitudes more solar radiation is received than lost. North and south of the lines more energy is lost than is received from the sun; thermal balance exists only because air and water heated nearer the equator flow toward the poles, carrying warmth. Astride these lines in the north sit Tokyo, Los Angeles, Gibraltar, and Tehran, Iran; and in the south are Buenos Aires,

Capetown, and Sydney, Australia. Here the year is a scant 306 thermal days long.

At 40 degrees from the equator people in Madrid, Peking, Philadelphia, Ankara, and Wellington, New Zealand, live with only 288.5 thermal days yearly. Along the lines of 50 degrees, which pass through Winnipeg, slide between London and Paris, lie just south of Irkutsk in Siberia, and touch the tip of South America, a year brings only 249.7 thermal days.

At 60 degrees north are Anchorage, Oslo, Helsinki, and Leningrad. In the south the imaginary line never touches land, but along it a year brings only 207.8 thermal days. At 70 degrees north, just south of Pt. Barrow, Alaska's northernmost tip, a year is 173 thermal days. At 80 degrees, on the Norwegian islands of Spitzbergen, a year yields but 156.6 days. And at the poles 90 degrees north and south, when the sun is below the horizon for roughly half of each year, the planet may spin around 365 times but the year is only 151.6 thermal days long. Far less than half as much solar energy falls over the poles each year than arrives over Quito, and this largely explains the difference in their climates.

Even the word "climate" comes from a Greek word meaning "slope," for the ancient Greeks believed the world sloped toward the North Pole, and that differences in warmth and weather between places were caused by this. They were essentially correct. Climatic zones on our planet are symmetrical to the equator.

In the United States, for example, according to former Yale University conservation professor Paul B. Sears, "on level land the growing season begins about twenty-four hours later for every fifteen miles one moves northward," a difference caused largely by diminished sunlight, hence less warmth. At these same latitudes, every one-quarter degree C. drop in temperature moves the "frost line," the boundary of the area affected by frost, 100 miles closer to the equator. Thus, changes in sunlight and temperature are equivalent to a change in latitude.

What if sunlight starts disappearing, as the NOAA scientists claim has occurred? If we subtracted ten minutes of daylight from a midwinter day in several American cities, the result would be tantamount to moving them north by several hundred miles. For example: New York City now receives no more sunlight than Boston did in 1950, and could be said to have "moved" to Boston.

In a like manner, Washington, D.C., has "moved" to the thermal-day latitude once enjoyed by New York City. Philadelphia has "moved" to Hartford, Connecticut. Chicago has "moved" to Green Bay, Wisconsin. Los Angeles has "moved" halfway to San Francisco. And Boston has "moved" into Maine. To the extent that heat from sunlight warms these cities, their climate will be influenced by this change. For the United States as a whole, unless this disappearing sunlight soon returns, the climate will gradually grow colder.

In 1971 Professor Budyko reported that recorded observations made since 1886 in the Soviet Union, North America, and Europe showed sunlight increasing its intensity until 1936, but decreasing continuously after that.

The twenty-three-year NOAA study completed in 1972 supports Budyko's findings. It recorded a decrease of autumn sunshine by 8 percent, offset somewhat by a 3 percent increase in springtime sunshine. One of the NOAA scientists, James K. Angell, described the disappearing daylight as an "interesting climatological phenomenon with impact on a local and perhaps hemispheric scale." Other nations, he said, should begin such research.

Why is the sunlight vanishing? Is the sun changing or is something in space interfering with its light? More likely, something is screening out the sunlight after it reaches Earth's atmosphere. But what?

Dust on the Window

When a window gets dirty, less light comes through it. Earth's atmosphere is like a window between us and the sun. According to Dr. Reid A. Bryson, Director of the Institute for Environmental Studies at the University of Wisconsin, the smoke and dust and other pollution humankind pours into the air have dirtied our window on the heavens. Thus, less sunlight can come through it.

All experts agree that certain kinds of dust at high altitudes can screen out sunshine and cool the air. Clear evidence of this comes from volcanic eruptions, which inject tons of particles high into the atmosphere.

Benjamin Franklin wrote that the cold winter of 1784 might

have been caused by volcanic ash in the sky. This seemed logical, for sky discoloration had long been attributed to distant volcanos. In 1783, Mt. Hekla, a volcano in Iceland, had erupted. In its wake sunsets were pink and red around the world, and at night the moon was blue.

In August 1883, the Indonesian island Krakatoa exploded with the force of many hydrogen bombs. The eruption sent clouds of dust, ash, and sulphate to a height of 50 miles, blotting out the sun for more than two days within a 50-mile radius and for nearly a day at an observation post 130 miles away. On globe-circling winds the particulate cloud soon reached France, thousands of miles distant, where the Montpellier Observatory reported a 20 percent drop in solar radiation. Sunshine was reduced by 10 percent there during the next three years, until the ash settled out of the atmosphere.

A cold winter followed the Krakatoa eruption, and cold winters also followed the 1783 eruptions of Asama in Japan and Mt. Hekla in Iceland, of Tamboro in the East Indies in 1815, and of Katmai in Alaska in 1912. The Alaskan volcano reduced sunlight by 20 percent in faraway Algeria.

In March 1963, a large volcano, Mt. Agung, erupted on the Indonesian island of Bali. Budyko reported that it caused a 5 percent loss of sunlight in the Soviet Union. Its worldwide effect by one estimate was a mean annual temperature drop of one-third degree C. that lasted two years.

Earth's volcanos have been increasingly active since the 1950s, but this change started after the cooling had begun. So why is the planet cooling? ''Man is so industrialized, urbanized, mechanized that he has become as important as natural phenomena in the modification of weather,'' said Dr. Bryson in 1966. ''The world's cities are putting out as much particulate matter as a volcano.'' And thus, according to Bryson, human pollution has the same effect as volcanic eruptions—putting dust on the atmospheric window, which reduces our sunshine. (Worth noting, many of the NOAA sunlight-measuring stations are in airports near major cities.)

Sunlight in the Soviet Union, writes Budyko, decreased by 4 percent between 1936 and 1960. He cites human pollution as the cause. Soviet researcher F. F. Davitaia studied the layers of dust frozen into Caucasus mountain glaciers. Dust increased until World War II, decreased when much of Soviet industry was destroyed, then increased again when new factories started polluting the atmosphere after the war; dust in the glacial layers increased 1900 percent between 1900-1960.

Sunlight began decreasing in the Northern Hemisphere at about the time of the 1930s Dust Bowl drought in the United States and similar droughts elsewhere. If Bryson is right, such dust could have helped cool the planet.

Climatologists used to think that such windblown dust, coming from drought lands or forest fires or mechanical farming, stayed near the ground and fell back to Earth in a few hours or days. Thus, it never should have influenced climate or produced more than a local effect. Today we know better. Since 1970 windblown dust from the drought-stricken lands south of the Sahara in Africa have regularly traveled on high-altitude winds across the Atlantic Ocean, blocking out up to 15 percent of the total sunlight reaching the surface of the tropical Atlantic. The dust in the skies of Barbados, an island in the Caribbean Sea thousands of miles from the Sahel, increased by 300 percent. While studying this, climatologists found that natives knew when desert tank battles were fought in Africa during World War II; days later the battle dust was visible over the Caribbean.

Almost all human dust and smoke settles out of the atmosphere quickly, within days or weeks. Much of what stays aloft longer than that is concentrated in Earth's far northern latitudes, where most of human industry is located. Thus, global measurements seem at first confusing. At the Mauna Loa Observatory in Hawaii, perched at 11,000 feet altitude atop an extinct volcano, measurements of atmospheric dust show it has increased little during the past decade. Soviet studies for dust effects between 1940 and 1967, however, show that in some places it now reduces sunshine by 10 percent. Measurements show that dust and smoke pollution over the North Atlantic Ocean nearly doubled between 1907 and 1970. Bryson often describes the "plume of pollution" he has seen "stretching from New York to Iceland."

At higher latitudes even noontime sunlight reaches the Earth at a steep angle. This angle means it must travel through more atmosphere. Where this atmosphere is full of smoke and dust, it must travel through more such pollution than it would to reach the surface at the equator. Thus, pollution dims sunlight most in the planet's higher latitudes, in Europe, the Soviet Union, and North America.

As industrialization increases, as more fuel is burned to make energy, more pollution will be created. As the need for farmland prompts more forest burning in the tropics, as mechanized farming spreads into new areas, as the global cooling causes more droughts, more such pollution will be created. A study completed in 1971 by Drs. S. I. Rasool and S. H. Schneider of the National Aeronautics and Space Administration's Goddard Institute for Space Studies estimates that man's potential to pollute will increase six- to eightfold in the next fifty years. This means humankind could increase the atmosphere's opacity by 400 percent within a century if pollution goes on increasing. That would reduce sunlight enough, say the scientists, to drop Earth's surface temperature by 3.5°C., which would almost certainly bring on an Ice Age. (Their calculations assume that dust reflects more solar energy back into space than it absorbs, but more recent data suggest this is inaccurate. In 1976 Dr. Schneider, now at NCAR, says atmospheric dust may absorb five times more energy than it backscatters, and this absorption could warm our climate.)

Some scientists even used to doubt that volcanos contributed to global cooling, but in 1975 a study of traces of volcanic dust in samples of sedimentary rock from ocean bottoms by University of Rhode Island oceanographers James P. Kennett and Robert C. Thunell produced evidence linking intense volcanic activity during the past two million years to Ice Ages. A large increase in atmospheric pollution can trigger cooling, these scientists suggest, and changes in sea level accompanying Ice Ages may trigger increased volcanic activity. Their preliminary findings suggest a positive feedback exists between volcanism and Ice Ages.

Human pollution, in quantity at least, compares well with the outpourings of volcanos. Over the United States in 1968, to single out one time and place, the National Air Pollution Control Administration estimates that 16.9 million metric tons of particulates and nearly 150 million metric tons of other chemicals and gases were dumped into the air. We use our skies as medieval Europeans used their streets—as sewers. We have yet to chart the precise effect such pollutants, singly or in combination, can have on sunlight in our atmosphere, but they undoubtedly absorb some sunlight and reflect some back into space, either of which reduces the light reaching Earth's surface.

Volcanos erupt once in a while, but human pollution goes on constantly. Many of the chemicals we put into the air today will remain aloft for five or ten years or more. And year by year such human pollution increases; if this has diminished sunlight in the past, we can expect less and less sunshine in the future. We could stop putting more dust on the atmospheric window by taking severe steps to stop pollution, but few seem willing to pay the price for doing this. We will let the window get dirtier.

The Great White Mirror

Of all the sunlight that reaches our planet, more than a third gets reflected back into space before it can warm us. Evidence suggests that in 1971-72 the amount of sunlight lost in this way in the Northern Hemisphere increased sharply. And sunlight reflected back into space is the same as sunlight never received.

Clouds can diminish sunlight, as we all know from the times a cloud has come between us and the sun on a summer day. Of the sunlight Earth loses to reflection, clouds cause more than two-thirds of the loss. Clouds also absorb a small percentage of the sunshine reaching our planet.

Cloudiness in our upper atmosphere has been increasing since 1950. High-level cloud cover over Europe, the North Atlantic, and North America has grown by 10 percent according to Dr. Bryson; this is a change he describes as "not negligible." The reason for the increase, he says, is manmade clouds, the jet contrails of aircraft along a transatlantic route so heavily traveled as to merit the name Sky Highway. Other scientists confirm that such cloudiness is increasing elsewhere, and that jet aircraft are a probable cause. More pollution could cause more clouds, too.

Down on Earth's surface everything is a mirror, reflecting light. But some things make better mirrors than others. The reflectivity of the planet, and of things on it, is called the *albedo*. The average albedo of clouds is 50 percent: they reflect back half the sunlight that hits them. Because at any given moment clouds are covering half the planet's surface, they reflect back into space about 25 percent of the total sunlight reaching Earth. But a calm ocean may absorb 98 percent of the noontime sunlight it receives and mirror back only 2 percent; hence its albedo is 2 percent. Land covered with grass or crops reflects between 8 and 25 percent of the light it gets. Forests, depending on the color of their leaves, reflect between 5 and 40 percent of sunlight. The albedo of deserts, as we know from the glare of sun on sand, is high; they reflect 30 to 35 percent of the sunlight they receive. And fresh snow has the highest albedo, bouncing back up to 85 percent of any sunlight falling on it.

Picture a green grassy valley, soft and warm, absorbing 85 percent of the sunlight it receives. Picture a sudden storm dumping snow over the valley. After the storm the valley can absorb only 15 percent of the sunlight that falls on it. Its albedo has risen from 15 to 85 percent. Such snowfall robs heat from the entire planet in several ways. The ground loses heat because it cannot absorb as much sunlight. If it had caught the light it would have converted it into heat, then given up that heat to the atmosphere. The sunlight arrived by traveling through Earth's atmosphere at a high rate of vibration, a short wavelength. Hitting the ground, it would have been changed to energy at a lower vibration rate, a longer wavelength called infrared light or heat. Our atmosphere can hold heat, much as the windows of a greenhouse* let visible light pass through but trap infrared light inside. But when sunlight bounces off snow, most of it is reflected back into space as visible light. It never becomes heat, and never gets converted into energy-binding forms of longer duration, such as green plants or that best-known product of ancient plants, fossil fuels like coal and oil.

Multiply this valley's area by millions of square kilometers and the impact of winter snow becomes apparent. Such an increase in the total albedo of the planet is, in effect, the same as a decrease in sunlight. Cooling is the consequence. According to George J. Kukla and Helena J. Kukla, albedo specialists with CLIMAP, the

---

*Some scientists dislike the term "greenhouse effect," preferring less colorful substitutes like "radiation-trapping effect." A greenhouse warms, they note, not so much because its walls are opaque to heat (infrared radiation) as because they reduce the convection that would otherwise carry heat away.

Climate Long-Range Investigation, Mapping, and Prediction program being conducted by four American universities, the location and duration of snow and pack ice fields are "the most important seasonal variable in the Earth's heat balance."

In 1974 the Kuklas announced results of their study, based on satellite photographs of ice and snow cover on the planet between 1967 and 1973. Their conclusion: the average annual snow and ice coverage in the Northern Hemisphere increased by 12 percent during 1971 and has stayed abnormally high thereafter. Winters (defined as snow and ice annual coverage in excess of 55 million square kilometers) grew almost a month longer, lasting 84 days in 1967 and 106 days in 1973. As of December 1975, the annual snow cover was still abnormally large but declining somewhat, according to studies by the National Oceanic and Atmospheric Administration. Most of this increase has been in Asia.

More snow cover means that our planet as a whole has a bigger albedo, that it reflects more sunlight back into space. Because sunlight lost is much the same as light never received, enough lost sunlight could mean serious cooling. Budyko estimates that a sustained increase in Earth's albedo of as little as 5 percent would be enough to start a new Ice Age. According to the Kuklas, in 1971 alone the yearly snow cover increased over what it had been the previous year by four million square kilometers. "Only seven similar occasions would be needed to establish the pleniglacial surface albedo," they wrote; i.e., snow coverage like that during the peak of the last Ice Age.

Even before 1971 the ice pack at the South Pole was expanding rapidly. Antarctic ice increased in summer area by about 10 percent between 1966 and 1970.

In the Northern Hemisphere during those same years the pattern of snowfall seemed as ominous as its growing volume. Its distribution, a Massachusetts Institute of Technology study noted, was remarkably parallel to ice distribution during the last Ice Age.

Seven Years From an Ice Age

Strictly speaking, we already live in an "Ice Age," a time when large parts of the planet are permanently covered with ice. But in popular usage, the words have taken on a special meaning: a period when ice has advanced toward the equator (or, more vividly, when snow endures year round in the places it now covers only in winter). During the last advance, which peaked 23,000 years ago, great sheets of ice covered much of Europe and the Soviet Union and crept south of Chicago and Seattle. Today, the snow fields are somewhat expanded, the Antarctic ice pack is expanding, and glaciers are advancing in many parts of the Northern Hemisphere. As long as the cooling continues, the ice can advance.

Weather scientists used to believe that such things happened slowly, but new evidence makes this view obsolete. Ice ages can come on quickly. Theoretically, we are only seven winters from another Great Ice Age, as the Kuklas suggest. We could possibly be less than a century from the start of one, warned the National Academy of Sciences in 1975. Such sudden cooling has happened on our planet before.

We know that 90,000 years ago the world was warm, only slightly cooler than it is today. But within a hundred years temperatures fell, plunging the Northern Hemisphere into a 1500-year period of extreme cold as severe as a Great Ice Age. Evidence of this comes from Greenland, from ice cores taken by Professor Willi Dansgaard of the University of Copenhagen. Where he drilled the snow had been built up layer by layer for hundreds of thousands of years. Some scientists believe these layers can be read as a climatic history of the Earth, each one a page recording the isotopic composition of the precipitation when it was formed. Dr. Dansgaard studies the amount of a special oxygen isotope, Oxygen-18, because it gives a clue to ancient temperatures. This heavy isotope, he assumes, evaporates from the oceans when temperatures are high. Thus, it occurs in the precipitation of warm eras. As climate cools, the proportion of Oxygen-18 in ice samples drops quickly, and during Great Ice Ages it becomes rare in the snow records. In the Greenland snow layers laid down 90,000 years ago, it suddenly became rare and it did not increase again for fifteen centuries, a clear indication of sudden and severe cooling.

Other evidence seems to support Dansgaard's estimates. Core samples taken from the ocean bottom in the Gulf of Mexico show that warm-water marine creatures disappeared in the region about

90,000 years ago and were replaced in "less than 350 years" by small creatures adapted to colder water. The cold-water creatures predominated in the Gulf of Mexico for 1400 years thereafter.

Dutch experts on fossil pollen believe that lush oak forests once flourished in what today are northern Greece and the Netherlands. They were devastated and died during a few brief centuries, 90,000 years ago. A rapid, intense cooling—a change of climate—could have killed them.

If Dansgaard's analysis is correct, we may be on the verge of another such cooling. The Oxygen-18 count was increasing until about 1930 in his study of snows in the twentieth century, but since then it has been falling rapidly—much as it did in Greenland 90 millennia ago.

An Ice Age can come quickly, but how? An ice pack is tremendously heavy, and according to one theory it can reach a critical mass. When that happens its weight creates enough pressure to melt the layer of ice at its base. Lubricated by this film of water, the ice pack can slide. It begins to surge outward and expand.

Glaciers surge this way, and have been recorded moving several hundred yards a day. But such speed typically lasts only a few days at most before progress slows, before the water lubricant is soaked up by the glacier's new territory and the stretching reduces the ice river's mass. As an explanation of sudden ice ages, this seems inadequate, and evidence that it happens with whole ice packs is scanty.

Another theory says snow packs can grow by momentum, by creating zones of cold air. The packs grow by expanding their snow boundaries a bit each year. This view could adequately explain slow-moving Ice Ages. The ice packs move like steamrollers, pushing back warm air in front of them and paving the way for snow to fall just beyond their edges. Fed by this snow, they grow. But this could not bring on an Ice Age within a century.

"Snowblitz" is the name used for the most popular current scientific theory of how ice ages can appear almost overnight. "Blitz" is used in the sense of *Blitzkrieg*, the German name for lightning war. Snowflakes land like a billion paratroopers, seizing land in winter. In summer they fail to melt completely. Less sunlight is absorbed by the land, and snow comes earlier to the cooled land the following winter. More snow remains the next summer, less heat is absorbed, and the land cools further. Frost begins blanketing bare parts of ground in late spring and early autumn, reducing warmth yet more. Within seven years the snow that falls does not melt. A thin coat of ice lies over the ground, reflecting away the sunshine that once warmed the land. The takeover is complete. Each year the snow thickens, gradually compressing into an ice sheet.

As Earth cools, as sunlight diminishes, as the range of snow and the length of winter increases, the possibility of such a snowblitz grows greater and greater. The winters of 1971 to 1973, when Mother Nature tried on her white mantle and sent shivers through her audience with a few cold gestures, may prove to have been a dress rehearsal for what the near future has in store. Ice and snow covers more area today than even a decade ago, and by some indications the cooling has only begun.

2

Waiting
for Warming

The cooling surprised weather scientists. Until it began, most had thought of climate change as something of the distant past or far future, something that happened slowly. Glaciers oozed toward the equator or retreated at a few miles per millennium. Hence the coming and going of ice ages seemed as uninteresting as the hour hand of a clock would be to a creature who was born, matured, and died in 10 seconds.

A few scientists at the turn of the century began to suggest that climate could change quickly, however slow past changes had been. The reason: the balance between heat and cold on the planet was delicate, easily tipped, and a new natural force on Earth was emerging. The force was humankind and its railroads, factories, steel mill blast furnaces, and internal combustion engines. More than a billion people inhabited the Earth in 1900, and the world was rapidly industrializing. All else being equal, these scientists said, human influence could create a new global climate—a warmer climate that might abolish ice ages forever.

Until 1950 these prophets of warming seemed correct. In the seven decades following 1880 the world's annual mean atmospheric temperature rose nearly 1°C. In Canada and the Soviet Union, the annual growing season increased by more than two weeks and crops could be raised more than one hundred miles closer to the North Pole than before. At the peak of the warming, between 1935 and 1942, a huge whirlpool 125 miles across formed in the Kuroshio Current off the Japanese coast, and for "seven years of plenty," fish harvests were large and rice could be grown in Japan's northernmost islands. Enough ice melted in Greenland and the Arctic to raise sea levels around the globe an average of four inches. If the warming continued, warned experts, the ice at both polar caps would melt and ocean levels would rise 400 feet, inundating many countries and all seacoasts on the planet. This would virtually re-enact Noah's Flood.

Then came the cooling, and within three decades global temperatures plunged more than halfway back to the 1880 average. Scientists were relieved but bewildered. They had carefully analyzed the reasons for the warming, so they thought, but the forces heating the world kept growing stronger after 1950, as climate abruptly cooled. Some have refused to believe in the cooling at all, dismissing the consistent measured drop in world temperatures (the longest ever recorded) as a temporary deviation in a longer-term warming.

"After the next decade or so will come a warming trend," predicted Dr. J. Murray Mitchell, a senior research climatologist with the National Oceanic and Atmospheric Administration (NOAA), in 1974. Many experts share his belief that the cooling will reverse, and that our climatic problem a century hence will be an overheated planet. They face a quandary: in theory Earth should still be getting warmer; in fact Earth has been cooling, and their theories cannot explain why. Perhaps less contradiction exists here than the fierce debates among climatologists indicate. Perhaps the warming has even *caused* the cooling. But before we can consider this we need to understand why some thoughtful scientists expect the climate soon to warm again.

The Global Greenhouse

At the birth of our solar system two infant planets were nearly identical twins. Earth and Venus had almost the same size and mass and chemical composition. The fate of the twins, however, has been far from identical.

Venus is now a greenhouse gone berserk. Her atmosphere blankets the planet with clouds of carbon dioxide gas and ice forty miles thick. The clouds act like windows on a greenhouse, trapping whatever solar energy reaches the planet's surface, which has an average temperature of 500°C. The atmosphere has no free oxygen and little water vapor. No life form we know could long survive there.

Why are these two planets different? Venus orbits 26 million miles nearer the sun than Earth does, so it receives much more solar radiation. But if Earth orbited only 6 million miles nearer the sun than it now does—a change of less than 7 percent from our 93 million-mile average orbital distance—our climate might become like that of Venus. Earth rotates once in 24 hours, Venus once in 243 days. Earth turns counterclockwise, and its creatures see the sun seem to rise in the East. Venus does the opposite, and if the sunrise could be seen through the dense atmosphere it would appear in the West. But rotational differences cannot explain why the two have such different climates.

Their difference lies in the amount of carbon dioxide in their atmospheres. On Earth CO_2 makes up roughly 330 parts per million of the air we breathe. On Venus CO_2 makes up more than 95 percent of the total atmosphere. This gas intensifies the greenhouse effect over both planets.

Venus gets far more sunlight than Earth, but this difference played its biggest role in changing both planets billions of years ago, in the evolution of the hydrogen-oxygen molecule H_2O, water. Most of the water that ever existed on Venus was evaporated, then, as a gas, and was destroyed in the planet's atmosphere by intense rays of sunlight. The resulting oxygen, scientists theorize, eventually recombined into CO_2. By comparison, 140 million miles from the sun on Mars, our smaller brother, water molecules formed but are now frozen in the cold environment. Only on Earth was sunlight of the right intensity to maintain H_2O in its liquid form. At present, of the water our planet possesses, 98 percent is liquid, 2 percent is in its solid form, ice, near the poles and in high mountains, and .001 percent is a gas in the atmosphere.

This liquid water saved us from the fate of Venus by protecting us from CO_2. Our twin planets have almost identical amounts of carbon dioxide, but on Earth it does not dominate the atmosphere.

Water helps lock up our CO_2 in several harmless and helpful ways. For one, CO_2 dissolves in water. Our oceans store perhaps sixty times more of it than is contained in the air. For another, life evolved in the oceans, and many life forms consume carbon dioxide. During Earth's Mesozoic Age (230 to 65 million years ago), volcanic activity created a huge increase in atmospheric CO_2, warming the planet and making an ideal environment for dinosaurs. But the new conditions also spawned oceans full of plankton, which learned to make limestone shells of the gas. Over millions of years the plankton depleted the atmosphere's carbon dioxide, which led to a cooling of world climate. The plankton ate themselves out of good conditions for survival, theorizes Dr. Thomas Worsley of the University of Washington, and cooled the planet enough to bring down the curtain on the age of dinosaurs. The shells of the planktonic organisms endure as limestone deposits throughout the world, trapping most of Earth's carbon dioxide in rock.

Green plants also devour CO_2 and use it in building their bodies. Each spring and summer in the Northern Hemisphere, as plants grow, carbon dioxide in the atmosphere drops by five parts per million. When winter comes the plants die and decay, returning the CO_2 to the air. But when the plants do not decay the gas is retained longer, as in wood and fossil fuels. In your home the wood in your doors, the plaster in your walls, the concrete in your foundation, and the plants you grow all are storing carbon dioxide and delaying its return to the atmosphere.

But this gas is merely locked away, not destroyed. It can escape by decay and burning. Animal bodies are slow furnaces that oxidize food. As you breathe while reading this, you release CO_2. Because this CO_2 was probably captured by other living things only recently, its release causes no harm. When you drive a car, however, or burn fossil fuels to heat your home, you are releasing carbon dioxide locked up in the bodies of plants millions of years ago when Earth's climate was tropical.

Human consumption of energy, most of which comes from fossil fuels, is increasing by about 5 percent per year. This increase is mirrored in the rising proportion of CO_2 in Earth's atmosphere. In 1900 the air was 290 parts per million carbon dioxide, but today is nearly 330 ppm. This proportion is increasing by at least two-tenths of a percent yearly, according to Soviet climatologist Mikhail Budyko, who fears that by year 2000 the atmosphere will contain up to 20 percent more CO_2 than it does today.

In 1970 scientists from around the world joined in a Massachusetts Institute of Technology forum for the Study of Critical

Environmental Problems (SCEP). They estimated that by the year 2000 carbon dioxide in the atmosphere would increase by 18 percent, to 379 ppm, and that the resulting greenhouse effect would raise world temperatures one-half degree Celsius. A doubling of atmospheric CO_2, they estimated, would raise Earth's mean annual surface temperature by 2°C., and this could happen within a century. By 1975 several of the scientists involved in SCEP had revised their estimates upward and said atmospheric carbon dioxide would reach between 390 and 400 ppm by the year 2000, thereby speeding warming all the more.

This trend, moreover, could gain momentum. How much CO_2 the oceans can absorb depends on temperature. Colder water holds more CO_2. But if climate begins warming, the seas will release part of the carbon dioxide now dissolved in them. In the atmosphere this extra CO_2 will increase the greenhouse effect, hence there will be more warming, and the oceans will yield up yet more CO_2. The consequence eventually could be the fate of Venus —what scientists call a "runaway greenhouse."

Without the greenhouse effect of water vapor and CO_2 to hold heat in our atmosphere, Earth would be at least 50°C. colder than it is today—a frozen planet without liquid water, without life. But with too much greenhouse effect the 2 percent of Earth's water locked in ice would become liquid and sea levels would rise drastically. By one estimate, the heating caused by a doubling of present CO_2 levels in the air would start melting the Antarctic ice cap at a rate that would raise ocean waters by 40 feet per century for the next one thousand years.

Other kinds of human pollution may warm Earth's climate too. Many kinds of dust from industrial smoke or engine emissions will hold heat in the atmosphere that otherwise would have gone back into space, according to NOAA's Dr. Mitchell and other scientists. But is the net effect of all dust in the atmosphere to warm the climate or to cool it? Scientists are as yet unsure.

One problem in figuring such things out is that some pollutants, if bombarded by intense sunlight in the upper atmosphere, change into new chemicals with unforeseen qualities. In 1974, for example, a controversy began over whether humankind was destroying all life on Earth with cans of hairspray and underarm deodorant.

The propellant most often used in such sprays is a class of gases called fluorocarbons, commercially sold under the DuPont trademark Freon. In 1973 University of California chemistry professor F. Sherwood Rowland found that ultraviolet light could break fluorocarbons down into other chemicals, including chlorine. Interesting, he thought, because few things in nature interact with fluorocarbons, which is why when released they tend to float slowly, unmolested, into the upper atmosphere, where solar ultraviolet breaks them down. By a fluke, however, Dr. Rowland learned of a government study that said chlorine would be very dangerous in the upper atmosphere, that it would set off chain reactions wherein one chlorine atom could destroy 10,000 ozone molecules. Thus the fluorocarbons could potentially devastate the layer of ozone floating more than 20 miles above Earth's surface.

Ozone is a special oxygen molecule, a trace compound in the atmosphere best known as the eye-sting in smog. Without it we would be dead. The ozone layer in Earth's upper atmosphere acts as a filter, absorbing only 2 percent of the sunlight reaching our planet but absorbing it from a select part of the solar energy spectrum—the ultraviolet light that comes at wavelengths too short for human eyes to see, too long to be called X-rays. We know what ultraviolet feels like; it causes sunburn. Without the ozone layer to shield us against the full intensity of the sun's ultraviolet light, virtually every species on Earth would eventually die. (Even creatures of the darkness must feed on things grown in sunlight.)

Rowland warned that fluorocarbons are damaging the ozone layer that protects us, and most scientists agreed. The subsequent controversy questions only how much ozone is being destroyed, how rapidly nature replenishes ozone in the atmosphere, and whether or not the risk merits a government ban on fluorocarbon sprays, a multibillion-dollar industry. By one projection, which assumes that the fluorocarbon industry keeps growing at 22 percent per year as it did between 1960 and 1972, the ozone layer will be depleted by 40 percent by 1995.

During the early 1970s public discussion about the hazards of the supersonic transport (SST), which would fly at high altitudes, led scientists to consider its effect on the ozone layer and to investigate ozone's relation to human well-being. In 1973 the National Academy of Sciences released a study showing that a 5 percent reduction in the ozone layer over the United States would mean a 26 percent increase in harmful ultraviolet radiation. A 50 percent depletion of the ozone, NAS reported, would increase ultraviolet radiation by 1000 percent.

What this means in human terms can be calculated in growing rates of skin cancer and other problems. A 5 percent drop in atmospheric ozone, NAS concluded, would cause roughly 8,000 cases of skin cancer among Caucasians (skin pigmentation offers significant protection). It would produce symptoms of aging: wrinkles, discoloration, dryness, and blood vessel dilation. Ultraviolet light damages DNA, the molecule in the body responsible for transmitting genetic information to future generations of our species; thus, like other kinds of radiation from atomic bombs, large doses of ultraviolet light might increase human mutation. It penetrates the skin, leading to effects not yet known in muscle and brain tissue, and in bone structure. The effects of a big ultraviolet increase on the ecosystem of the planet is also little understood, but some experts believe it could kill most plankton on the surface of the world's oceans, thereby radically disrupting food chains and other natural relationships.

By one estimate, the fluorocarbons already sprayed into our atmosphere will cause 150,000 skin cancers and 6,000 deaths per year for the next decade, even if all use of the chemical stopped today. The ozone layer will not be naturally restored for at least a century.

Still more subtle effects can be considered. Many types of plants studied, including tomatoes, lettuce, peas, and millet, show a retardation of growth when the proportion of ultraviolet in their light increases. This could limit their ability to hold CO_2 and hasten their release of it through earlier death. In marginal areas, such as desert edges and mountain timberlines, an increase in ultraviolet rays might stop plant growth altogether.

In terms of climate, more ultraviolet radiation means more solar energy striking Earth's surface. This could cause snow, and perhaps even polar ice, to melt.

In 1975 the Climatic Impact Committee of the National Research Council recommended quick action to protect Earth's ozone. Committee chairman Henry Booker, a University of California scientist, said that unless limits were put on SST aircraft and the use of fluorocarbon propellant gas, the ozone layer might decrease by 60 percent by the year 2000, with "disastrous consequences for human health, agriculture, and climate."

By the end of 1975 few such actions had been taken. Oregon imposed a law banning fluorocarbon use after 1977, but no other state followed suit. Ten states have petitioned the Consumer Product Safety Commission to ban such sprays and have been turned down. The final federal report on the problem was slated for April 1976, after which action might be taken. Use of the sprays continues, although many companies have voluntarily stopped making fluorocarbon products or offered alternative products to consumers. American interest in manufacturing its own SST has waned, but in 1975 the British-French Concorde SST toured U.S. airfields and was ready to begin international passenger service, and the Soviet Union began regular domestic cargo flights with its own TU-144 SST.

Heat Pollution

According to a natural rule we call the Second Law of Thermodynamics, Earth should be warming. This is because one of the implications of that law is that all the energy human beings use is eventually degraded to heat.

Quietly sitting, reading a book, a human being gives off 450 British Thermal Units (BTU)* of heat each hour. A 4000-pound automobile driven at seventy miles per hour gives off 750,000 BTU. The refining of one barrel of crude oil to make gasoline for that automobile produces 150,000 BTU of heat. Every bit of energy we use adds heat to the global environment, and we are using more and more energy.

In 1925, total human energy consumption was an estimated 44 quadrillion BTU, but by 1968 it had jumped to 190 quadrillion BTU. Our world energy use has been growing by more than 5 percent per year. At that rate, by 1980 we will be consuming 345 quadrillion BTU annually. By the year 2000 that will increase to 831 quadrillion BTU, almost a nineteenfold increase in seventy-five years.

This added heat will have an immediate impact on Earth's climate, wrote Soviet climatologist Mikhail Budyko in 1974. Ice

*A BTU is the heat needed to warm one pound of water one degree Fahrenheit.

is melting in the Arctic already. By the year 2000, he expects, "the average boundary of polar ice will retreat northward about two degrees" of latitude. "Distinct global changes in the climate will occur when the ice floes in the Arctic are completely melted —which will take approximately 80 years."

Few scientists agree with Budyko's timetable, but many share his belief in the warming. In 1973, for example, research meteorologist James T. Peterson of the U.S. Environmental Protection Agency projected that if energy use in the world keeps growing by about 5 percent annually, in year 2075—less than a century from now—humans will be putting 1 percent as much heat into the environment as the sun does. This, he said, would almost certainly "lead to significant climatic changes."

What would the climate be like? For clues, consider some of the world's bigger cities, all of which are "heat islands" in which human energy production causes lots of heat. Already in wintertime New York City generates two and one-half times more heat than sunlight delivers naturally. Already the heat humans create in Chicago and Washington, D.C., gives those cities frost-free growing seasons a full month longer than surrounding rural areas (a virtue offset by smog, which reduces crop yields as much as 33 percent).

By the year 2000, experts estimate, the urban megalopolis of Bosnywash—the continuous city covering 11,000 square miles from Boston through New York to Washington, D.C.—will be pouring out manmade heat equal to 50 percent of the sun's heat in winter, 15 percent in summer.

The climate will be warm, as anyone knows who lives in a big city—from 1 to 3 degrees C. warmer than if the city were not there. Asphalt streets and concrete pavements soak up sunlight. Buildings compress heat over those streets, creating zones of still air sometimes warm as an oven. The city cools little at night, for the streets keep radiating the heat they catch in daylight. The city is warmer than the countryside in winter, too, for more snow melts while falling through warm city air and when hitting warm city streets.

Each year a bit more of Earth's surface becomes city. Year by year human energy consumption, and hence human-created heat, increases by about 5 percent. At present the demand for more energy is expanding the ratio of heat to power: the old fossil fuel plants that generate electricity are typically 40 percent efficient, creating four units of electricity for every ten units of coal or gas energy they consume; but nuclear power plants are typically 30 percent efficient and thus put much more heat into the environment than would a fossil fuel plant producing the same amount of electricity. The newest reactors are more efficient, but the moral they and fossil fuel facilities and geothermal plants all offer is the same—one price of more energy is more heat pollution of the planet. One price of more heat pollution, sooner or later, will be a warming of Earth's climate.

Four sources of energy add no heat to the environment: solar, hydroelectric, wind, and tidal-current power. These tap energies already loose and generating heat in the world. But apart from the 1975 commitment of the U.S. government to develop solar power, little will be done with these options in the near future. The power of falling water is already largely exploited. Wind is generally deemed too unreliable for large-scale power needs. Ocean currents and tides have great promise, but huge investments of capital would be needed to tap them. For the near future almost all power will be generated by burning fossil fuels or nuclear fuels.

We will be releasing more heat on our planet, and from many of these same burning processes we will be releasing more carbon dioxide into the atmosphere. The CO_2 will intensify Earth's greenhouse effect, which will hold more of the heat in our environment that otherwise would have escaped into outer space. How can this do anything but warm Earth's climate? If you like the tropics, you will like the future—unless you live within 400 feet of sea level, in which case you should learn to swim, build an ark, or begin looking for a home on higher ground. Thus say the prophets of warming climate.

Such arguments seem sensible, logical, almost compelling. But if they are accurate, why has Earth's climate been cooling rather than warming since the 1940s? One answer is that they fail to take all the factors into account. As University of Wisconsin climatologist Dr. Reid Bryson points out, most of the human activities that release carbon dioxide and heat into the environment also cause dust or other particulate matter in the atmosphere. Sometime around 1950 the climatic effect of dust and increased cloudiness may have overwhelmed what until then was a strong global warming trend. Will heat and CO_2 eventually overtake the present cooling? Not likely, says Bryson, because the forces causing both heating and cooling have, in most cases, the same sources.

Human vanity is such that we like to claim credit for all changes on the planet, whether we cause them or not. Climate swung through warmings and coolings for millions of years before humankind arrived here, and it will go on doing so long after we depart. Our power to change such things is as yet tiny compared to nature's. Take energy, for example. As Bryson notes, since 1950 the North Atlantic Ocean surface has cooled one degree Celsius. What would be needed to restore this temperature drop, equivalent to 15,000,000,000,000 watts of power? If we could channel all human energy, from the oxcart driver in India to the atomic reactors of America, into this one task, it would take nearly two years just to raise the temperature of one small part of one ocean by $1°C$. If all the waste heat from all the power plants projected to exist on Earth by the year 2000 were dispersed evenly in all the oceans, average sea temperature would increase only $.00001°C$.

The day may come when we can heat and cool oceans as we now do swimming pools and office buildings, but that day is not here yet. A computer model designed for ocean studies and used by the National Center for Atmospheric Research suggests that human beings could generate forty times the heat they do today without significantly changing our climate, but this is little more than a widely-debated sophisticated guess; we are only beginning to understand the gigantic forces of heating and cooling in nature.

What of carbon dioxide and the greenhouse effect? It is over-rated, said Drs. S. I. Rasool and S. H. Schneider of the National Aeronautics and Space Administration Goddard Institute for Space Studies in 1971. Temperatures do not increase in proportion to an atmospheric increase in CO_2. Beyond a factor of two to four, the effect of carbon dioxide levels off. Even an eightfold increase over present levels might warm Earth's surface less than $2°C$., they wrote in *Science*, and this is highly unlikely in the next several thousand years. Venus cannot be compared to Earth, for its greenhouse literally has a window, a layer of CO_2 dry ice, in its atmosphere.

If the amount of energy we use keeps growing, the resulting heat would eventually influence climate—but in complex ways, perhaps by making additional clouds that would block sunlight. In the United States in 1974, electrical energy consumption *decreased* by a fraction of a percent rather than increasing by 5 percent—so the timetable some doomsayers suggest for a melting of Earth's polar caps may require readjustment.

No final scientific judgment has been rendered concerning depletion of our planet's ozone layer, but some points should be noted. Above your head the ozone layer varies naturally up to 25 percent each day and an additional 25 percent each year. At eleven year intervals ozone in the upper atmosphere increases by 5 percent, which suggests it may be influenced by the sunspot cycle. Since 1955 the ozone layer's density has *increased* by 8 percent in the Northern Hemisphere and decreased slightly in the Southern Hemisphere. Since 1970 a 2 percent decrease has been measured in the Northern Hemisphere, but this still means ozone has increased 6 percent since widespread fluorocarbon use began.

Certainly, disappearance of the ozone layer would be catastrophic. Whole species might die or mutate. Moreover, if our atmosphere contained ten to twenty times the fluorocarbon gases it does today (one-tenth part per billion), Earth's climate would probably warm enough to melt the polar ice caps and flood many cities—not so much because of additional solar radiation because of a thinner ozone shield as because chlorofluorcarbons increase the atmosphere's greenhouse effect, according to calculations by Dr. Veerabhadran Ramanathan of NASA's Langley Research Center. (Other scientists say a fluorocarbon increase would *cool* our climate by reducing the total greenhouse effect ozone provides.)

But this is unlikely. Ozone is a renewable resource, rapidly replenished in our atmosphere. Production and use of fluorocarbon gases is now in sharp decline. Other threats—ranging from the release of ozone-destroying nitrous oxide from nitrogen fertilizers to the stratospheric pollution of the SST—should be carefully monitored and studied, but present evidence indicates that panic is unwarranted.

As Tom Alexander, an editor of *Fortune* magazine, has written, "it might even turn out . . . that there's too *much* ozone up there and that everybody will have to get busy with his aerosol cans again.'' It may, indeed, prove out that the ozone layer increase, combined with the increase in atmospheric dust, may have absorbed a critical amount of sunlight and contributed to the cooling in Earth's Northern Hemisphere.

Earth's climate has been cooling. This fact seems to contradict theories that say it should be warming. But the prophets of warming *are* describing real forces that influence climate, and like other scientists they are still learning how these forces interact to produce a balance of heating and cooling on our planet. It may well turn out that the growing instability of Earth's climate is caused by human influences adding both heating *and* cooling forces to the balance, thereby making it more and more ''unnatural'' and precarious. The prophets of both warming and cooling agree on at least one thing: climatic changes can come quickly, within centuries or even decades, and can have devastating consequences for humankind. Climatology has ceased to be a drab science. Its findings have taken on an urgent importance for all of us.

3

Climate in
the Balance

They came to the Canadian Arctic in 1960 to study the cooling of
Earth's climate, never imagining what they would witness.
From then until 1969, Drs. R. S. Bradley and G. H. Miller,
scientists from the University of Colorado, carefully watched the
changes in snow and temperature on Baffin Island, north of Hudson Bay. In 1972 they reported their bizarre findings.

As expected in a cooling climate, snow and ice increased on the
huge island. Old glaciers grew bigger, and two new glaciers were
born. But something was strange: the annual mean temperature on
Baffin Island, which lies across the Arctic Circle, *was unchanged.*
Through the year temperatures varied. In June, July, and August
the thermometer fell 2.1°C. below normal, but during the remaining nine months of each year temperatures averaged 2.0°C. *above*
normal. Averaged over whole years, temperature on Baffin Island
remained the same, even though "the landscape seems to be
moving toward more glacial conditions." They accepted an obvious conclusion: that ice can increase even without a significant
drop in temperatures. Perhaps the same is true of the advance of an
ice age.

In 1934 the head of Britain's Royal Meteorological Office, Sir
George Simpson, proposed an amazing theory. Ice Ages, he
wrote, are caused not by cooling of the planet, but by heating. The
theory seems quite logical when we pause to consider it. Ice needs
the same thing to grow that green plants do: water. But the polar
regions of the Earth are deserts: frozen lands, they are covered with
heavy masses of cold air which block warmer moisture-bearing

winds coming from the tropics. Thus, the Arctic and Antarctic have fierce storms but little precipitation; their ice endures but does not grow.

Simpson believed that the sun varied, and that when it for a time put out more solar energy, the Earth's equator warmed much faster than the poles. Evaporation of sea water increased in the tropics, and water-laden winds pushed toward the poles with added vigor. Reaching near the poles, this moisture fell as precipitation—as snow, or as rain that froze to ice when the sun brought summer to the other side of the planet. This ice and snow chilled the air above, and reflected sunlight that the land beneath it had once absorbed. The ice and snow expanded, fed by warmer, moist winds from the tropics, and out of equatorial heating an Ice Age was born.

Some evidence seems to confirm Simpson's theory. According to British climatologist Hubert Lamb, head of climatic research at the University of East Anglia, the surface of the North Atlantic Ocean warmed slightly between 1970 and 1974. This, say Lamb and fellow researchers, accounts for the good winters Europe enjoyed during these years. Conversely, during these same years snow cover in the Northern Hemisphere reportedly *increased* by an amazing 12 percent. This could be a clear instance of heat causing cooling. Studies by A. T. Wilson of New Zealand's University of Waikato and C. H. Hendy of Columbia University's Lamont-Doherty Geological Observatory show that during Ice Ages the temperature difference between equatorial and polar regions has increased 20 to 25 percent and, as fossil sand dunes in Australia confirm, winds moving poleward were apparently stronger.

Other evidence questions his theory. The polar regions cooled more rapidly than areas near the equator during past Ice Ages, but the equatorial zones gradually cooled too. Thus, total evaporation should have decreased, even if winds were stronger. This happened slowly, perhaps more slowly than Simpson's analysis allowed. But he replied that this was untrue of the Pacific Ocean, which was cut off from Arctic cold waters when falling sea levels closed the Bering Strait. Thus, said Simpson, the Pacific became a warm pond and had a high evaporation rate. Studies of ocean bottom cores now indicate he was wrong, although evidence cannot rule out the possibility of a very brief warming whose evaporation might have kicked off an advance of the ice.

And how could the poles grow colder because more heat reached them? Simpson's defenders might reply that the polar regions, where temperatures today can dip below −73°C., have an "abundance of cold," enough to freeze many times the available moisture. Thus, when more moisture arrives, the cold is more than sufficient to offset heat brought by wind, and create snow. This snow spreads cold on the planet, but this causes not so much a change in Earth's thermal balance as a redistribution of energies—even as a seasonal redistribution of energies on Baffin Island caused glacial conditions without changing the average annual temperature. During Ice Ages, say some of Simpson's defenders, the polar regions grow somewhat warmer, and large parts of the Arctic Ocean may become free of ice.

Simpson's theory has many flaws and may well be wrong, as the most evolved computer models of climate now seem to indicate, but it helps suggest the complexity of climate on our planet.

Is the Earth cooling or warming? The correct answer is, always has been, always will be, BOTH. Our planet takes in vast amounts of heat energy from the sun and re-radiates a nearly identical amount to outer space. This gives us what weather scientists call "thermal balance." It must be so. If solar energy diminished by only a small percentage, our planet would quickly lose heat and freeze. If, by greenhouse effect or for some other reason, we retained very much more energy than we now lose, the polar ice would soon melt, the world would flood, and after a time the seas would boil away into vapor, giving us a mantle of clouds like Venus'.

The devices by which our planet maintains thermal balance are, in their detail, extremely subtle. In general terms, however, the mechanism commonly called Earth's "weather machine" is simple and straightforward. It is a giant heat pump, gathering warmth near the equator and distributing it by liquid and gas working fluids which circulate between the equator and the poles.

Water is the first, and most important, of these fluids. The oceans contain a thousand times more heat than Earth's air, and ocean currents deliver this heat throughout the oceans, even under the Arctic polar ice. Ocean temperatures change slowly, and in summer remain cooler than the air above the oceans and in winter warmer. Thus, the oceans act as a climatic "flywheel," helping moderate temperature throughout the year.

Air is the second fluid. It carries less heat than the oceans, but it moves much faster. It also distributes fresh water, as precipitation, throughout the world; as vapor, this water stores most of the heat found in the winds. The air helps control climate both over the oceans and over land where the ocean never reaches.

Why do the air and oceans move? One reason is our planet's rotation. At the equator the Earth is 25,000 miles around, and in twenty-four hours it turns around once. Thus, if you stood at the equator you would be moving nearly 1,100 miles per hour because of Earth's rotation. If you stood at the North Pole, on the other hand, you would be traveling zero miles per hour, because the poles are the points around which Earth turns. At different distances between the equator and the poles, this speed is different, and it imparts different energies to fluids on and above the surface of the planet. A glance at the way ocean currents and regular winds move, as shown in Figures 3-1 and 3-2, suggests the influence of Earth's rotation.

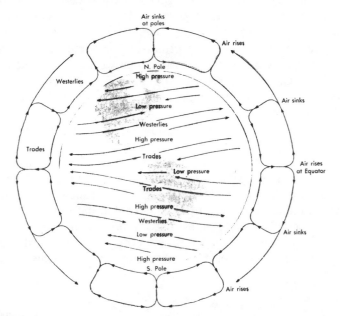

FIGURE 3-1

Simplified schematic drawing of the general circulation of the atmosphere.

Source: Louis J. Battan, Weather (Englewood Cliffs, New Jersey: Prentice-Hall, 1974).

FIGURE 3-2
Simplified drawing of the general circulation of the oceans.

A second reason for their motion is thermodynamic law, which holds that in this universe heat will disperse itself toward cold (cold being merely the absence of heat, not a thing unto itself) with vigor proportional to the difference between the heat at the beginning of the cycle and the cold at its end. This is why modern power plants use steam "superheated" to around 260°C. rather than steam near the 100°C. boiling point of water—the hotter the steam, the more work output it yields in transforming from hot to cold points in the cycle. In the same way, water and wind heated by sunlight near the equator have a natural tendency to move toward colder parts of the planet near the poles.

Conversely, when the winds and waters cool they lose energy. Near the poles ocean currents grow denser as they cool and sink to the ocean bottom, where they drift slowly back toward the equator. Lighter, warmer currents flow poleward over the top of these cold bottom currents. What remains of their vigor drifts down the west coasts of the continents, in latitudes far enough from the equator as to be little affected by the Earth's eastward rotation; when near enough to the equator these waters will warm again and begin moving west, beginning the cycle again.

At several points in their poleward motion, moisture-bearing winds are forced downward by cooling and drop part of their vapor as rain. A map of the world's deserts would clearly show where the biggest of these downflows occurs, along the edge of the westerly winds, for when these winds touch ground, their ability to deliver rain is small, having just been spent in the subtropics. Deserts neatly rim the lands of Earth to the north of the Tropic of Cancer, in North Africa, the Middle East, China, and the American Southwest. Exceptions to this desert belt occur only where mountains intervene, or where huge warm bodies of water provide extra heat and evaporation immediately equator-ward. Mountains and land masses play key roles in determining how much water the winds carry to certain nations, and whether those winds blow with or without obstruction. This shapes the climates of various regions.

By analogy we might say that global weather and climate change when the world grows new "mountains" or tears down old ones—not mountains of rock, but of cold air and cold water, both of which have significant density. Around the North Pole, for example, lies a huge mass of cold air called the "circumpolar vortex" that effectively blocks most of the warm winds moving north. Likewise, in the North Pacific, North Atlantic, and Arctic Oceans masses of dense cold water deflect warm currents flowing north. During an Ice Age these cold air and water masses push closer and closer to the equator, and this prevents warmth from reaching the far northern latitudes of the planet. Thus, the North grows colder, and this coldness helps extend the cold water and air farther south. The Ice Age grows.

At present, the Gulf Stream current carries warm water from the tropics north along the East Coast of the United States, across the North Atlantic, and against the shores of Northern Europe. During the last Great Ice Age, dense cold waters in the North Atlantic deflected the Gulf Stream south, dumping its cooled current against Morocco in North Africa. Deprived of Gulf Stream heat, Northern Europe was soon buried under snow and ice. Ocean temperatures off Spain resembled those near Greenland today.

Along the southern edge of the northern circumpolar vortex zips a fast air current called the jet stream, flowing around the planet like the hem of a twirling skirt. Where it dips south, cold polar air follows, chilling the weather. Where it swerves north, warm winds from the south can come. Either way, zones of air pressure are created that reinforce the stream's path, so once a pattern is established it can last years, and potentially decades or longer. British climatologist Hubert H. Lamb speculates that the "Little Ice Age" that devastated Northern Europe from about 1400 until 1850 occurred because the jet stream developed a long-term "dip" that incessantly poured cold polar air over Europe, and he says the cold land and air of the continent and the North Atlantic reinforced this pattern for centuries.

During the 1960s, for a present-day example, a stationary high-pressure area developed over Greenland. This caused a slight vacuum effect in wind patterns that increased cold winds sweeping down into Europe across the Norwegian-Greenland sea. As a result, Europe suffered some severely cold winters. In 1970-71 this pressure zone collapsed. Heavy snows followed in the Northern Hemisphere, especially in Asia; but since then winters in Europe have been milder. Sea ice has retreated somewhat in the North Atlantic, and ocean surface temperatures have warmed in a few spots. Some of the critics who had long argued that the global cooling never existed rushed into print saying the nonexistent climate change "was over." Many thought the summer of 1975 proved them right. A heat wave blistered across Europe. Factories in Sweden shut down for five days as temperatures soared above 95°F. How could Earth's climate be cooling?

But as Dr. Lamb pointed out calmly, such heat waves have accompanied every past global cooling and are to be expected. A high-pressure zone blocked warm air and chilled the North Atlantic. Now another high-pressure zone was blocking cold air and bringing extremes of heat into Europe. But such blocks were both symptoms of a cooling climate. Such cooling, he said, "means more volatile weather. It will be more hot, more cold, more wet and more dry, just as it was in the seventeenth century."

Such pressure blocks work somewhat like mountains in the sky, but built of air pressure rather than rock. Whether their short-term effect is to warm an area or cool it, their longer-term effect is to retard global wind patterns in the far northern latitudes, and thereby to reduce the total amount of heat reaching the North Polar region from the equator.

The Gulf Stream has shifted slightly south since 1950 and shows no sign of shifting north again. The cod have moved with the current, and Iceland has gunboat patrols at sea trying to keep foreign vessels outside her newly declared 200-mile fishing limit in the Atlantic Ocean. Within a few years, Dr. Lamb made clear, the same pressure blocks in the winds that warmed Europe in 1975 would relocate, and cold waves would follow the heat waves. Every global cooling is marked with wild extremes in temperature and precipitation.

But, as Lamb realizes, people tend only to believe in what happens at the moment. When a pressure cell develops in the Earth's wind patterns, it soon reinforces itself by what weather scientists call "thermal forcing." A zone of cold air over an ocean, for example, helps cool the water surface beneath it. The cooled ocean then continues interacting with the air, cooling and deflecting any warmer winds that pass over. Such interactions, which Dr. Jerome Namias of the Scripps Institution of Oceanography in San Diego, California, sees as the biggest force in shaping prolonged weather changes, mean that mild or harsh weather can continue for years over some regions, even when that weather is contrary to global trends. Brief periods of good or bad weather can thus fool people about what is happening to world climate.

During the 1960s, analyzes Namias, warm regions of ocean surface in the North Pacific brought good weather to the western United States by changing the zigzag pattern in the jet stream, but the northward zig over the western U.S. prompted a southward zag over the eastern section, and thus eastern winters were unusually cold. Today this is changing. Patches of warm and cold water in the North Atlantic created a record number of icebergs in 1971 and 1972, and the "thermal forcing" they caused bent the jet stream again. The eastern United States and Europe began to benefit from northward zigs in the skirt of the circumpolar vortex. But in Soviet grainlands the warm air brought little moisture, only drought, and in 1972 and again in 1975 harvests suffered. The zone of low pressure created over the southeastern United States sucked in Hurricane Agnes, which killed more than 100 persons and did $3 billion in damage. Winters were mild.

But on America's West Coast, temperatures slid downward. In southern California the 1975 spring was one of the coldest since the 1880s. Autumn and winter, 1975–76, were the driest in a century.

The reason for such odd extremes of weather becomes clear when we look at Figure 3-3, which shows how the airborne "mountains" of the circumpolar vortex have changed since 1950. In general, the cold air from the North Pole has been reaching farther south in recent years than it has since the nineteenth century. This figure shows an averaging of the sharper dips of the jet stream. Because cold air reaches farther south, the possibility of cooling is evident. Because warm air can settle over areas nearer the poles, the possibility of warming is evident. On balance, however, the shift in global winds favors global cooling over the long term for several reasons. The bulk of the circumpolar vortex is moving south. As it moves, world wind patterns in the North will be disrupted, and for every bit of area warmed by high-pressure cells, a larger area will be cooled because of this retarding effect on normal winds. The tracks of storms moving from the Arctic will move farther south, bringing rain and snow and tor-

FIGURE 3-3 *Simplified map of changes in global air mass boundaries.*

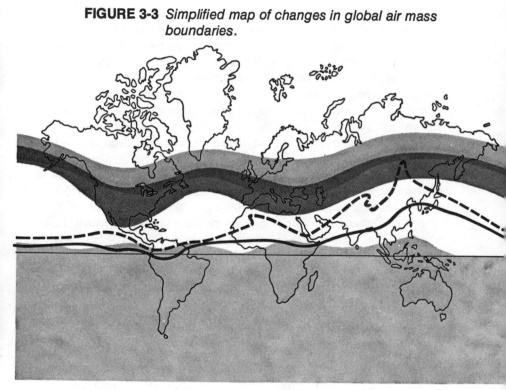

Key:

Position of the Circumpolar Vortex Boundary 1900–1940

Position of the Circumpolar Vortex Boundary in 1970

=== Position of the Intertropical Convergence Zone (Northern Limit of the Monsoons) 1900–1940

— Position of the Intertropical Convergence Zone (Northern Limit of the Monsoons) in 1970

Portion of the Earth experiencing climatic warming. These same areas experienced cooling from 1900 until 1940.

nado-like disturbances. Finally, as suggested by Sir George Simpson's theory, these far-ranging high-pressure zones serve to carry moisture into a cooling Arctic region where it will become ice and snow, as happened over the Northern Hemisphere when the jet stream shifted between 1970 and 1972.

Nearer the equator, another region of global wind pressure reacts to changes in the circumpolar vortex. This is called the Intertropical Convergence Zone. It is the northern limit of the monsoon rains on which more than one quarter of humankind relies for crop irrigation. As the circumpolar vortex shifted south, the Intertropical Convergence zone did likewise. This, many experts agree, caused the drought in the African Sahel and Ethiopia from 1968 through 1974, the dry years that came in India during the same period, and the lack of rainfall in parts of the Caribbean and Central America. Optimists hailed the scant rainfall in the Sahel in 1974, and the good rains in India and the Caribbean in 1975, as evidence that the monsoons have returned to normal. Much as we would rejoice if this were true, it seems unlikely. Adequate rain in a few nations is probably the result of slight temporary changes in the Intertropical Convergence Zone in response to brief new zigs and zags in the jet stream. The cooling continues, but even if it were warming, the monsoons would probably not return to normal in this century, warn specialists such as University of Wisconsin climatologist Dr. Reid Bryson. At present the overall wind pattern of our planet appears to be returning to the way it was during the colder climate of a century ago, and this helps explain the many newspaper stories saying we are experiencing the wettest, driest, coldest, hottest conditions "in over seventy years," or "since 1880," or the like.

Lately, those who say the global cooling is over or never existed (or both) have pointed to two other facts to bolster their case. One is that the region of Earth's equator is warming rather than cooling. The other is that while the Northern Hemisphere cools, the Southern Hemisphere has been heating up. But these phenomena, too, could be products of a global cooling.

As the cold air mass of the North Pole has pushed south, less heat has reached the Arctic. What happens to this heat? Some has been manifested in snow and ice. Of the remainder, some is deflected back toward the equator. The lower latitudes of the planet are now containing the same amount of heat that was spread over a wider area, so of course they are hotter. And if air and water get warmer around the equator, and no comparable force of cooling blocks their motion toward the South Pole, we can expect some of this increased heat to spill southward. This apparently has happened.

What of a greenhouse effect in our atmosphere caused by growing concentrations of carbon dioxide gas—which experts say should heat our whole planet? And what of dust in the atmosphere, which other experts say should cool us? ''Respected atmospheric scientists have recently warned me that we could be getting the worst of both effects,'' said Canada's Deputy Minister for the Environment Robert F. Shaw in 1974, because ''particulate matter tends to go to the poles, which means the temperate zone countries will be cooled further while the equatorial zones receive the carbon dioxide. This may be causing what seems to be a shift in the monsoons.'' Shaw did not name these scientists, whose views are debatable. Dust, moreover, can act on the atmosphere in complex ways that intensify wind blockages.

The Southern Hemisphere has very little industry and hence little dust. It also has amazingly little land, because oceans predominate south of the equator, and oceans soak up solar energy much more effectively than does land. Moreover, these oceans can circulate more freely than can oceans in the land-heavy Northern Hemisphere, so snow has little chance to settle on land if it should fall. The whole present Ice Epoch on the planet has centered on the North Pole, with cooling only reaching southward during each Ice Age after ice was far advanced in the north. Thus, such reputable scientists as M. J. Salinger and J. M. Gunn of the University of Otago in New Zealand are right to remind us that climate is warming at present in the Southern Hemisphere. However, they are surprisingly lax in their failure to mention that the Antarctic ice pack is increasing in area, just as would be predicted by Sir George Simpson's theory. Even as the air retains or increases warmth, the ice spreads—just as on Baffin Island. In New Zealand the mountains since 1970 have received what one scientist calls the heaviest snowfall in many years. In tropical Brazil the coffee-growing regions have repeatedly experienced severe frost and snow.

Meanwhile, the cooling in the Northern Hemisphere alone has

been powerful enough to reduce the average annual mean air temperature of our *whole* planet by one-half degree Celsius, according to available measurements. Our world is cooling faster in the North than it is heating in the South, and more than 80 percent of humankind lives north of the equator. In several important ways, the thermal balance on our planet seems to be tipping toward further cooling, and quite possibly toward an ice age. If an ice age comes, within a few centuries the majority of surviving human beings may live in the Southern Hemisphere. At the height of the last Great Ice Age many parts of Australia were cool, rainy, and rather pleasant when compared with most other lands on Earth.

4

Father Sun,
Mother Earth

We depend on the sun for almost all energy on Earth. Its light warms our planet, drives the winds, raises the rain clouds, and makes all life possible through photosynthesis in plants. But when climate cools and when measurements show that sunlight is diminishing, we must wonder: is the sun dependable, or could it be changing?

Scientists believe the amount of sunlight we receive is unwavering, and is one of the most reliable forces in nature. They call this energy rate "the solar constant" and define it as the radiation reaching the edge of Earth's atmosphere at right angles when at our "mean [annual] distance" from the sun: 1.94 gram-calories per square centimeter per minute. American science writer D. S. Halacy, Jr., translates this as roughly 1000 watts of solar power per square yard per minute, equivalent to about one and one-third horsepower.

Funny thing about this "constant"—it is not constant. Many factors can and do change it. And Earth's climate could be drastically altered, some experts say, by tiny variations in the vast amount of solar energy we receive each year—variations as small as two calories per square centimeter *per day*, a loss of two minutes' worth of sunlight. Such changes are too small to be detected by present measurement technology.

These things we know: the intensity of light striking any object decreases with the square of the distance between the object and the light source. The distance between Earth and the sun varies. During its annual journey around the sun, Earth presently travels 3 million miles closer to the sun in January than in July. This means that the sunshine we in the twentieth century receive is weaker by 7 percent during the northern summer than during summer in the Southern Hemisphere.

Every 10,000 years this situation reverses, because Earth, spinning like a top around its axis once every twenty-four hours, also wobbles like a top, tipping on its base once every 21,000 years. In the year 12,000 A.D. the northern half of our planet will be tilted toward the sun when our annual orbit comes closest to it; then the Northern Hemisphere will receive 7 percent more sunshine. Astronomers call this wobble-cycle the precession of the equinoxes.

Also like a top whirling in space, Earth changes the angle of its axis relative to the sun. During a cycle that takes roughly 40,000 years, our planet tilts from 21.8 degrees off perpendicular in the solar plane to 24.4 degrees, then back again. The smaller our tilt, the smaller the difference in temperature that should exist between winter and summer. For the past 10,000 years the angle of our tilt has been getting smaller, which in theory should make summers slightly cooler, winters slightly warmer. Such ''lowered seasonal contrast'' favors cooler climate. More snow tends to fall on warm days than on cold, natives in snow country know, because snow depends on evaporated moisture in the air, which depends on some warmth. Cooler summers, however, mean that snow is slow to melt and soon to return. As increased snow cover reflects back more and more sunlight that once was absorbed, cooling intensifies. Wind blockages form in the higher latitudes of the planet, and this retards the global circulation of warm winds from the tropics—creating odd warm spells in some places, strange unseasonable cold in others. A warm winter, therefore, may be one symptom of climatic cooling.

At present, Earth's orbit around the sun is nearly circular. But during a cycle which astronomers say takes 90,000 to 100,000 years, Earth's orbit stretches into a long ellipse and then returns to a circular shape. When our orbit is stretched, the intensity of sunlight reaching our planet varies by as much as 30 percent during a single year, a change that inevitably would cause climate change.

One of the most puzzling things about Great Ice Ages is their regularity: they come roughly once every 100,000 years. Beginning in the 1920s, a Serbian geophysicist, Milutin Milankovitch,

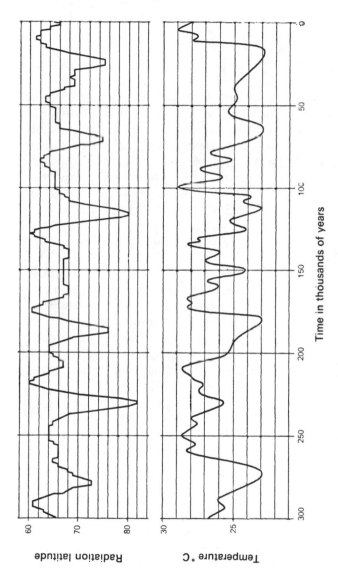

FIGURE 4-1 *Solar Variation and Ice Ages*

This 1958 chart was done by marine fossil expert Dr. Cesare Emiliani. In it he compares ocean temperature changes for the past 300,000 years, based on his analysis of sea bed sediments, with summer solar energy changes predicted for latitude 65 degrees north (the latitude of Iceland) during the same period by the Milankovitch theory. The solar energy variation is expressed as a change in latitude. "Thus 25,000 years ago," he wrote, "latitude 65 received no more solar heat in summer than latitude 75 does today."

Source: Cesare Emiliani, "Ancient Temperatures," *Scientific American* (February 1958), p. 61.

advanced a theory that such climate changes were caused by Earth's movements relative to the sun. Without the help of a computer, he spent years correlating the overlapping effects of our tilt, wobble, and changing orbit. At first his work was ignored by other scientists. The popular dogma held that Earth had had only four Great Ice Ages, spaced at odd intervals; Milankovitch's theory fit poorly with this.

But by 1955 new evidence was appearing. Measurements of the heavy oxygen content of layers in the ocean floors showed that Earth has had at least seven Great Ice Ages, coming at remarkably even intervals, as Milankovitch's theory predicted. A leader in the ocean-floor research, Dr. Cesare Emiliani, compared his findings with Milankovitch's calculations and found the two amazingly similar, as shown in Figure 4-1. "The main difficulty with the Milankovitch theory," wrote Emiliani, "is that it fails to explain why the Ice Epoch developed only recently—within the last million years—after 200 million years during which the Earth had no ice ages." One possible reason for this, suggested Emiliani, was the emergence of more land from the oceans. New continents limited the circulation of pole-bound ocean currents, and new mountains slowed pole-bound winds. Before the new land rose, wind and water had circulated heat freely throughout the planet. The land, moreover, could retain snow that would melt if it fell on water. And the land would absorb less sunlight than did the waters it pushed aside. For all these reasons an increase in land on the planet would encourage cooling.

The bulk of recent scientific investigations seem to support Milankovitch's theory, if not his own crude calculations. In 1975 British science writer Nigel Calder compared the current picture of Earth's climatic history, derived from elaborate studies of ocean-floor samples, with the most sophisticated computer study yet done on astronomical changes in the sunlight Earth has received over the ages. The result, shown in Figure 4-2, is a refined version of Emiliani's chart. The correspondence it shows between changes in sunlight and Earth's climate is clear.

If the Milankovitch theory is correct, as evidence now seems to suggest, then humankind has acquired more than a tool for understanding the past. Because we can predict precisely how Earth will

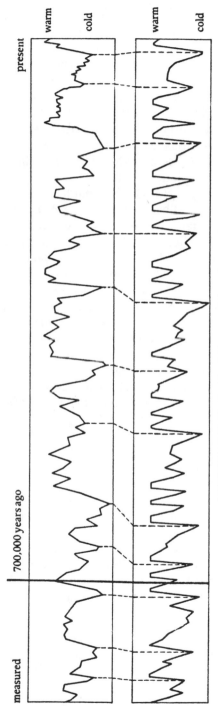

FIGURE 4-2 *Solar Variation and Ice Ages*

This 1975 chart was done by British science writer Nigel Calder. In it he compares ocean temperature changes for the past 800,000 years, based on analysis of sea-bed sediments by Drs. Nicholas Shackleton and Neil Opdyke, with summer solar energy changes predicted for latitude 50 degrees north (the latitude between London and Paris) as calculated using Milankovitch theory by Dr. Anandu Vernekar of the University of Maryland. The line at 700,000 years marks a major reversal in Earth's magnetism that aided scientists in coordinating their statistics. Calder assumes that whenever sunlight is two percent stronger than at present, ice diminishes. When less than that, as today, ice accumulates in proportion to the shortfall at .22 the rate of melting.

Source: Nigel Calder, *The Weather Machine* (New York: Viking Press, 1975), p. 132.

move relative to the sun in the future, we gain a tool for prophecy as well. The prophecy it gives us, however, is chilling.

"If the theory is correct," wrote Emiliani in 1958, "about 10,000 years from now there will be another advance of the glaciers, burying Chicago, Berlin and Moscow under thousands of feet of ice. Indeed, we can expect periodic glaciations to continue until the Earth's great mountain chains have been weathered down to hills." His prediction, based largely on the influence of the 40,000-year Earth-tilt cycle, anticipated that summers were subtly cooling and that this would cool the planet. When he wrote, fellow scientists generally ignored his opinion. They had begun to believe that Great Ice Ages happen every 100,000 years, but the last one peaked only 23,000 years ago. And between these Ice Ages, the common belief held, came uninterrupted periods of warmth.

By 1972, research by Dr. Emiliani, now at the University of Miami, and by a growing number of other scientists, had made the comfortable older dogmas obsolete. The normal temperature for Earth's climate, they found, is much colder than what we are now experiencing. During the past 700,000 years, Emiliani estimates, global mean temperatures have been as high as in this century only 5 percent of the time. Temperatures tend to peak and dip every 20,000 years, roughly the time it takes Earth to pivot once around its axis, and many shorter temperature oscillations take place within this cycle. Earth started its slow climb back to warmth from the frozen heart of the last Great Ice Age approximately 17,000 years ago.

The warm peak of the present interglacial period happened in about 6000 B.C. Since then we have been in a slow, bouncy downhill spiral. By the Milankovitch theory, according to Calder's comparison of the most recent data and computer projections, Earth will continue to grow colder without any major change in the present cooling trend for the next 27,000 years. Then, temperatures will level off, more or less for 20,000 years, in a range much colder than today's climate. Forty thousand years from now it will get colder and remain so for 80,000 years, causing perhaps the longest Great Ice Age in Earth's history. But in 120,000 years warming will come, and climate will again resemble what we have today.

Through all this time there will be warm years and cool, wet years and dry. But the general trend of climate will be toward cooling for the next 120 millennia. The thermal balance of our planet is tipping downward.

Among those who agree with the Milankovitch theory that astronomical motions shape our climate, there are several different predictions concerning our immediate future. Many agree with George J. Kukla of Columbia University's Lamont-Doherty Geological Observatory, whose recent studies show that sunlight reached its lowest late summer and autumn intensity 17,000 years ago, during the coldest period of the last Great Ice Age, and rose to its greatest late summer and autumn intensity more than 6,000 years ago, during the warmest period of the present interglacial. His calculations indicate that late summer and autumn sunshine will go on diminishing for at least the next 4,000 years. He believes we may today be at the dawn of a new Ice Age.

Drs. Emiliani, Kukla, and others emphasize that Earth's changes in relation to the sun can certainly change our climate. All else being equal, the Milankovitch theory explains the bigger variations in our planet's climatic history. But new factors can enter the picture: new mountains, volcanos, the position of Earth's magnetic poles, and most recently human pollution—all may influence the amount of the solar energy we receive that will be retained in our biosphere.

And other smaller cycles, recognized by climatologists, also influence our warmth. Every twenty-six months the circulation of westerly winds over the tropics goes through a cycle. Sunspots rise and fall in a double cycle averaging 11 and 22 years and cause a longer 80-year cycle of warming and cooling in global climate. A cycle of 18.6 years, caused by complex gravity interactions among the Earth, moon, and sun, influence geyser activity in Yellowstone National Park. Cycles, the origin of which is less understood, also occur at intervals of 180, 400, 800, and 2,500 years. These cycles overlap in odd ways, sometimes reinforcing and sometimes negating each other, and deciphering their complex influences has given many scientists work to do.

Earth's climate should soon begin warming, wrote climatologist Wallace Broecker of the Lamont-Doherty Geological Observatory in 1975, because we have just passed the cool forty-year phase of the eighty-year cycle. He could have added that we also passed the cold peak of the 2500-year cycle during the "Little Ice Age" between 1430 and 1850.

But as the panel of experts of the National Academy of Sciences warned in 1975, we are now entering the cold 10,000-year phase of the 20,000-year Earth-wobble cycle, and unlike many lesser cycles that vary our global temperature by less than one degree C., this giant cycle has always in the past plunged our average global temperature at least 3 degrees C., potentially enough to cause an Ice Age.* "The question naturally arises," said the NAS report, "as to whether we are indeed on the brink of a [10,000-year] period of colder climate." This seemed likely, they said, because our present warm climate is "highly abnormal," the sort of warm interval that in the past has lasted only a short time and then ended suddenly.

When we take these cycles into account, plus the Milankovitch prediction of cooling, plus the fact that our climate has already been briskly cooling, plus the possibility that present trends can gain momentum from various natural feedback mechanisms, we know as sensible gamblers that the odds favor more cooling in the near future. What are the chances that a Great Ice Age will begin within 100 years? One in ten, said scientists questioned by Calder in 1974. Thus, to ignore the possibility is to play Russian Roulette with a ten-chambered revolver.

The Changing Sun

The year 1974 was not good for food growers. In the American Midwest spring floods were followed by summer drought. Food production in the region fell by 16 percent, 6 percent more than what the U.S. Department of Agriculture officially calls a "disaster." It was the worst growing season since the 1930s, many farmers said. Similarly, odd weather devastated crops in Siberia, India, Canada, and the African Sahel. In most parts of the world harvests were poor. "Weather," said a National Science Foundation journal, "was a crucial factor."

Some meteorologists have called the 1974 weather disaster the "Walter Orr Roberts Memorial Drought," naming it after the

*NAS charts of climate cycles appear in Appendix I.

former director of the National Center for Atmospheric Research who predicted its coming more than a year in advance. Dr. Roberts sees nothing mysterious in his prophecy. "There have been something like eight successive serious dry periods [in the high plains of America's Midwest] spaced approximately twenty to twenty-three years apart," said Roberts in a March 1974 interview in *U.S. News & World Report*. "If the drought doesn't come within two more years or possibly three, then I'll feel that the cycle has been broken for some reason." In February 1976, Dr. Roberts predicted the then-existing midwest drought would last until 1981.

Like a growing number of scientists, Dr. Roberts has turned to the sun to learn about Earth's weather. The natural cycle most resembling the drought cycle in America's Midwest is that of spots on the face of the sun. On average, every eleven years the number of sunspots rises to a peak, then declines. The most recent cycle peaked in 1968 to 1969, fell off, and started to rise again in 1973. Cycles can vary by several years; the next peak is expected in around 1980.

In the Northern Hemisphere, wheat harvests tend to be good during peak sunspot years and poor during the minimum years of cycle. Dr. Joe W. King of the Appleton Laboratory in Slough, England, finds that Scotland's growing season is longest in peak years, and perhaps the same is true elsewhere. King found that in the Northern Hemisphere rain and sunspot increases follow precisely the same cycles at 50- to 60-degree and at 70- to 80-degree latitude, but that these cycles run exactly opposite one another at 60- to 70-degree north latitude. Dr. King suspects that the sunspot cycle influences the density of the ozone layer in Earth's atmosphere, which varies widely with latitude, and that through this influence sunspots change the amount of radiation reaching our planet's surface. In the Southern Hemisphere, harvests are best during the minima of the cycle, and generally poor when the sunspot cycle peaks. Australian scientist E. G. Bowen reports that rainfall diminishes in his country during peak sunspot years— except on the island of Tasmania, where rain greatly increases.

With each new cycle the total number of these spots on the sun's face changes. The peak number began decreasing around 1880. In about 1928 the cycle peaked with the lowest number of sunspots (fewer than eighty) in a century or more. This was the last peak, according to Soviet climatologist Mikhail Budyko's data, before sunshine began to diminish in the Northern Hemisphere. The number began to increase thereafter with each cycle, and as it did the Scottish growing seasons lengthened proportionately. By 1958 'tne cycle peak showed a near-record number of sunspots, nearly two hundred. Earth's climate began warming in the 1880s, reached its greatest warmth in the 1930s, and started to decline as the number of sunspots increased. Is there a connection? To find out, we must consider what sunspots are, and how the Earth is connected to the sun.

Sunspots appear darker than the rest of the sun's surface because they are cooler than it is by 40 percent. They are hot, nevertheless—about 3,800° Kelvin.* They have magnetic fields each at least 1,000 times stronger than Earth's and the whole sun's. One theory holds that sunspots are magnetic lines of force from inside the sun that have become rolled up like balls of yarn and have floated to the surface. They appear in pairs or small clusters, usually near the sun's equator. These pairs or groups have their own polarity: one spot will be positive and lead the motion of the negative spots across the solar face for one sunspot cycle, but when the spots fade and the next cycle begins, the spots re-emerge with this polarity reversed. Thus a full sunspot cycle, from plus to plus polarity, lasts twenty-two years, the length of time between droughts in the U.S. Midwest. On opposite sides of the sun's equator the spots travel in opposite directions and show opposite polarity, all of which reverse with each new cycle.

The sun is a great outpouring of energy, but where these spots sit with their strong magnetic fields, that flow can be bottled up for a time. Perhaps this explains the solar flares—huge eruptions of solar cosmic ray particles, X-rays, radio waves, and brilliant white light—that occasionally burst forth in the vicinity of sunspots. These flares throw powerful bursts of energy against Earth and other planets of our solar system.

Why are these flares and sunspots important to us? Because Earth is *inside* the sun, floating like a cork within the sun's atmosphere. We are surrounded not by a "vacuum of space," but

*Kelvin is identical to Celsius, but starts from absolute zero rather than water's freezing point: Kelvin = Celsius minus 273°.

by solar plasma, a fine gas made up of charged ions, atoms stripped of their electrons. Energy from the sun ripples out through this plasma as a "solar wind," buffeting all it touches, and disturbances on the sun produce interplanetary storms in this plasma atmosphere that can readily interact with weather on Earth.

Because it is made of charged ions, plasma is a good conductor of magnetism and electricity. Sunspots produce intense magnetic fields, as do flares, and the waves of ions they throw into space carry strong magnetic fields and electrical energy toward Earth at speeds up to 1,100 kilometers per second. Within a day after a solar flare erupts, or sunspots put out an extreme pulse of energy, such waves strike the Earth with tremendous force. One common result is the Aurora Borealis, the dancing Northern Lights above the North Polar regions, as this energy lights up gases over our planet's magnetic pole in precisely the way electricity lights a neon bulb. The Aurora is strongest when sunspots are at a maximum, and on rare occasions can be seen as far south as Florida.

Could changes in this solar wind influence Earth's weather? Certainly, speculated Dr. Roberts in 1967: ions in the solar wind would change ionization in our upper atmosphere, and "some subtle change of the stratospheric ionization may, in selected terrestrial locations, cause cirrus cloud banks to form. These, in turn, may change the radiation balance of the atmosphere enough to provide substantial heating or cooling . . . [hence] weather changes."

Radio amateurs have long known and waited for peaks in sunspot cycles, for the increase they cause in atmospheric ionization can make the atmosphere act like a mirror for long-range radio signals on shortwave bands. But flares and sunspots also cause distortions in layers of the atmosphere and in Earth's magnetic field that create static across the radio frequency spectrum—these are called electromagnetic or geomagnetic storms.

In 1973 Roberts and National Oceanic and Atmospheric Administration scientist Roger Olson announced their discovery that the same geomagnetic storms that cause the Aurora also change weather over the Gulf of Alaska, where winter storms begin that travel south into the central United States. Following geomagnetic storms, large low-pressure troughs tend to form over the far North

Pacific Ocean. These help push storms deeper into the continental U.S., and make those storms colder.

In 1974 meteorologists Harold Stolov of the City University of New York and Ralph Shapiro of the Air Force Cambridge Research Laboratory announced another connection between solar and Earth weather. Changes on the sun disturb Earth's magnetic field, and as a result the westerly flow of winds in our planet's mid-latitudes increases by up to 7 percent, more than enough to influence weather patterns.

Earlier that same year, Dr. King of the Appleton Laboratory published his controversial theory that Earth's climate is shaped by changes in our planet's magnetic field. The pattern of air pressure throughout the world, he found, corresponds closely with the pattern of magnetic intensity on Earth's surface. This, he speculated, may be because our magnetic field influences the intensity with which the solar wind strikes different regions of our upper atmosphere. The Little Ice Age that gripped Europe between the sixteenth and nineteenth centuries may have been caused by our ever-changing magnetic field, he said, for in 1660 the magnetic declination at London and Paris was zero—far different from what it is today—and this might have opened Europe to icy winds from the North Pole.

The reason is the jet stream, which ripples at high speed around the North Pole's cold air mass. Where the jet stream veers south, cold air is free to advance; where it moves north, cold air is somewhat fenced in nearer the pole. Today two spots of extreme magnetic intensity exist in the Northern Hemisphere: one in the Canadian Arctic, and one in Eastern Siberia. The jet stream always dips south of these spots, as if directed by them. Perhaps in 1660 a similar spot was above London and Paris, causing their climate to be somewhat like Siberia's.

The sun sits in space, a giant unshielded nuclear reactor. If nothing stood between us and it, we would die as surely as if we faced a small atomic explosion nearby on the planet. Two things protect us. Our atmosphere absorbs large portions of dangerous radiation, such as the ultraviolet light that is taken in by our ozone layer. And far from the Earth, our magnetic field deflects charged particles and rays like a huge shield and draws them to our magnetic poles, where they make the lights of the Aurora.

During intense magnetic storms our shield, called the magnetosphere, takes a pounding, and is sometimes compressed to half its normal size by the onslaught of particles, by the blast of the solar wind. As the magnetosphere buckles, increased amounts of radiation hit our atmosphere. Radio transmissions are disrupted. The Northern Lights flicker more brightly. The magnetic field of our whole planet reacts.

Humankind has been able to duplicate some of these effects. In 1962, for example, the U.S. Defense Department carried out Project Starfish, which involved the explosion of a large nuclear device 250 miles above Earth. The blast generated an aurora, destroyed some atmospheric ozone, and tore a temporary hole in Earth's protective ionosphere. (The launch of Skylab in May 1973 tore a similar ionosphere hole like "that created by magnetic storms.") It also caused radio and radar signal blackouts across part of the upper atmosphere; military planners worry to this day that an enemy attack might neutralize U.S. radar by such blasts in a war. The U.S. government has never released full results of Starfish, so we cannot determine other effects it might have had on weather, climate, or human health. Soviet research using Kirlian photography to study bioplasmic electrical fields, or "auras," around living things has found that these fields flare up brightly in response to flares on the surface of the sun. This suggests that the electromagnetic and other-energy fields set up by the sun and by atomic explosions might have a direct influence on living things. The sun, as astrologers joke, is the star under which we all are born, and it certainly must shape our lives. Could sunspots cause "sunacy" as the full moon was said to cause lunacy? Nobody yet knows, but research at the Douglas Hospital in Montreal provides evidence that the violent mass outbursts of patients in mental wards throughout the world correspond with sunspot activity. And the possible influence of sunspots on society "is not to be lightly dismissed," wrote economist John Maynard Keynes, who believed harvests rose and fell with the sunspot cycle—and that this affected the business cycle. In many parts of the world harvests vary by 10 percent, and in some areas by 50 percent, in accord with the sunspot cycle; and a peak in the solar flux in 1936 corresponded with world depression, but the connection between sunspots and world economics is unproven.

In August 1972 one of the largest solar flares ever observed erupted. It sent tidal waves of particles crashing against Earth. Two astrophysicists, John Gribbin and Stephen Plagemann, wrote that this flare's impact had slowed our planet's rotation by a few thousandths of a second, and that the U.S. Naval Observatory's Universal Time measurements showed this. They were widely criticized. Scientists at England's Royal Greenwich Observatory compared data from six stations in the network that measures Universal Time and found that the time changes after the flare were "certainly nothing more remarkable than occurred several times in the year. . . ."

Gribbin and Plagemann, their credentials as prophets seriously in doubt, published an even more startling work in 1974, *The Jupiter Effect.* In it they defended their evidence of the flare's impact on Earth, and expanded on the sunspot theories of University of Colorado scientist K. D. Wood.

What causes sunspots? Dr. Wood in 1972 suggested that perhaps the planets do. We know that both the sun and the moon cause "tides" in Earth's atmosphere, both by heat and magnetism. But the sun is a great glowing bag of superheated gas, and it seems possible that certain planets—particularly the three closest ones, Earth and Venus and Mercury, and the biggest, Jupiter, which is one-tenth the sun's size—could cause tides on the sun with their own magnetic pull. Jupiter travels around the sun every 11.86 years, which means that once in every average sunspot cycle it orbits the sun. The sun is slightly tipped, at about seven degrees, to the plane of these four planets, so once in 11.86 years Jupiter comes closest to the solar north and south poles. If this causes sunspots, why are their cycles sometimes irregular, varying from eight to fourteen years? The answer, says Wood, is that the planets work in combination. Earth orbits once yearly. Venus marches around the sun every 225 days, and Mercury makes the trip in eighty-eight days. Sunspots come, said Wood, when the gravitational pull of two or more of these planets are either aligned or at right angles to one another.

What fascinated Gribbin and Plagemann about this controversial theory was a calculation dashed off by Dr. Wood—that once every 179 years all the planets of our solar system cross the same line from the sun at the same time. During this time they exert tremendous tidal gravitational pull on the sun, and on each other Such an alignment of planets will next happen in 1982.

The consequence of this conjunction, wrote Gribbin and Plagemann, will be more than sunspots or flares. Great stresses will occur in the Earth's crust, and along the San Andreas Fault in California giant earthquakes will destroy San Francisco and Los Angeles. Critics again attacked, saying that no such earthquakes happened during the last alignment 179 years earlier, and that alignments nearly as intense happen frequently without any effect. In the revised paperback edition of *The Jupiter Effect* issued in 1975, the authors claimed to have found evidence of minor earthquakes in San Diego and Santa Barbara, California, that took place in 1800. The present alignment of planets, they wrote, will begin to take force in 1977.

Could a convergence of planets cause massive earthquakes? Perhaps, but certainly it has had other influences that Gribbin and Plagemann overlooked. A study of Greenland ice cores by University of Copenhagen scientist Willi Dansgaard, for example, reveals several regular cycles of cooling on our planet. One happens every eighty years, and the other *every 180 years!* This second cycle corresponds well with the planetary alignment Dr. Wood calculated, but nobody had connected the two before, including Gribbin and Plagemann. The last cooling in this cycle began 185 years ago and lasted for roughly 75 years, probably through self-sustaining feedback mechanisms. This cycle could help explain the present global climatic cooling.

As one raised in California with earthquakes rocking my cradle, I find lack of imagination among the greatest sins in the Gribbin-Plagemann thesis. Let us suppose that 1982 brings the fierce interplanetary gravity taffy-pull they anticipate. Earthquakes certainly might come. Recorded deaths from earthquakes in recent centuries suggest that the earth shakes most violently in December and January, when Earth comes nearest the sun on its annual journey—and folk-wisdom has always said that earthquakes happen during eclipses, when the sun, moon, and earth are aligned —so the possibility of planetary alignments causing quakes cannot be dismissed. Such quakes would devastate more than California. The whole mountain chain from Chile to Alaska would be affected. Along this chain are several dormant volcanos—such as Mt. Baker in Washington State, that could erupt at the slightest provocation—and the Jupiter Effect could set these off. At least twelve such volcanos could erupt on the U.S. West Coast alone, spewing toxic fumes and ash high into the air, where it would block sunlight and cool global climate. Other earthquakes would strike Japan, Hawaii, Central and South America, the Red Sea region, and the Persian Gulf (where oil wells might be rendered inoperative). Italy, Greece, Turkey, and Pakistan would suffer massive fatalities from earth tremors, and more deaths from disrupted water and food supplies.

But, if one wants to fantasize a scenario of terror, this is but the beginning.

In recent times Earth's magnetic field has been decreasing by about 5 percent per century. This is natural. Over tens of thousands of years our planet's magnetic poles diminish until, at some point, they "fall" into the core of the planet, only to re-emerge years or decades later at opposite ends of the planet. The North Pole becomes South and vice versa. By itself, this would probably mean little more than a name change. However, the process might also influence climate. In 1973 a research team from the Lamont-Doherty Geological Laboratory headed by Dr. Goesta Wollin concluded that Earth's climate tends to be warmer when and where the magnetic field is declining. As the field weakens, so does our planet's magnetic shield in space; perhaps more solar radiation gets through on wavelengths that can warm us—as opposed to those Dr. Roberts studied, which becloud the stratosphere. Science is still unsure why sinking magnetism seems to correspond in the short term to warming climate, at least in the past.

But what happens when our magnetic poles fall into Earth's core? The answer: our magnetosphere shield collapses, and only our atmosphere is left to protect us from solar and interstellar atomic radiation. According to Canadian geophysicist Robert Uffen, this may help explain why dinosaurs were suddenly replaced in global dominance by furry mammals, which in turn soon may be replaced by insects. During the past 600 million years our magnetic field has reversed at least 171 times, and during each of these reversals Earth was bombarded for a time by heavy doses of solar radiation. Like other atomic radiation, this produced vast numbers of mutations and probably helped kill off many radiation-vulnerable species. Magnetic reversals thus have been a major force in the evolution of species. The last brief (2000-year) temporary reversal happened 30,000 years ago, at the beginning of an Ice Age it may have helped cause. The radiation this poured onto our planet doubtless played a part in the evolution of early man too, and helped make us what we are today.

Are we on the verge of another magnetic field collapse, another dose of mutating radiation? According to Dr. Neil Opdyke of the Lamont-Doherty Geological Laboratory, to judge by the average of time between great polar reversals in the past we are 200,000 years overdue for one. When it comes, the ionosphere and magnetosphere will disappear for a time, and cosmic rays and assorted particles will bombard Earth at will.

Why would it come? According to Opdyke's colleague, Dr. Bruce Heezen, any severe shock to the planet could trigger it —such as a collision with a comet or giant meteor, which probably caused the last great magnetic reversal 700,000 years ago, and also probably scattered glass tektites across our world. Short of such a shock, if the present magnetic decline continues, our poles will collapse into Earth's core in about 2,000 years.

Could a series of nuclear weapons exploding cause such a collapse? This is unlikely. What of the Jupiter Effect, the periodic alignment of planets Drs. Gribbin and Plagemann expect in 1982? This too is unlikely, especially in light of new calculations by Belgian astronomer Jean Meeus released in 1975 showing that in 1982 Jupiter, Saturn, Uranus, and Neptune will *not* be aligned, contrary to the Gribbin-Plagemann prediction, and that those planets which *do* align will put a tidal pull on the sun 2.7 million times smaller than what the moon exerts on the Earth's tides. Could such a tiny pull cause a flip in our magnetic field? The possibility cannot wholly be ruled out. The only issue is the relative weakness of the field. At some point it will become vulnerable even to comparatively mild shocks and stresses, such as the 179-year cycle of planet alignments might cause. The Jupiter Effect may not collapse Earth's magnetic field in 1982, but sometime between now and the year 4000 it could do the trick. When this happens, as it inevitably will sooner or later, Earth will be dosed for months, perhaps years, perhaps decades, with high-intensity radiation. Some species will be destroyed, others will survive in mutated form, others will divide into several new subspecies. (We will die from our own nuclear radiation—because global radio communications will be washed out by static when the ionosphere collapses and the ensuing panic in thousands of missile and bomber bases will doubtless precipitate a world war.) These imaginary scenarios are much more exciting than mere California earthquakes, and would make good fodder for Hollywood movies, of late so eager to cash in on cataclysmic themes.

But the odds are that none of these things will happen in our lifetimes. Our concern is more immediate and probable: what do sunspot patterns suggest our weather will be between now and the year 2000? Will the global climatic cooling continue? Dr. Wood predicts, on the basis of solar tides caused by future conjunctions of the planets, that the next two sunspot cycles will peak in 1982 and 1993, and that both will produce fewer total spots than in the peak passed in 1969. However, in this computation he seems to have forgotten his own 179-year planetary alignment, the tides of which may cause a sunspot *increase* between 1980 and 1982.

Why do sunspot cycles seem to affect the Northern and Southern Hemispheres differently? Various possibilities exist. One is that sunspot impact is the same on both sides of the planet—that during peaks solar radiation increases somewhat on some wavelengths and decreases on others—but the two hemispheres are themselves different. A glance at a map reveals that little land is to be found south of the equator; oceans predominate, and oceans absorb heat better than land and refuse to retain snow as land will. During the global cooling trend in the Northern Hemisphere, the world south of the equator has been heating up. This heat can manifest itself as drought or flood, depending on circumstances. More radiation thus might help plants north of the equator while harming crops in the Southern Hemisphere.

Another possibility is that the plasma from sunspots, which carries with it a strong magnetic field, is drawn to one of Earth's poles more than the other. Evidence to support this theory is scarce. In 1960 University of Minnesota scientist Edward P. Ney reported that as the sunspot cycle moves toward a minimum, the number of cosmic rays striking Earth reaches a maximum, and vice versa. He suggested that this cycle opposite the sunspot cycle involved "positively charged cosmic rays" thrown off by the sun as its electrical potential grew in intensity. Could these cosmic rays be drawn to one of our magnetic poles, or be repelled by the other? The cosmic ray cycle remains a mystery.

Suppose the pattern of sunspot cycles in recent decades continued more or less unchanged into the immediate future. What could we expect? Consider the pattern: In 1958 the cycle peaked, and weather was good in the Northern Hemisphere. When the cycle was at minimum in 1962 to 1963, one of the worst winters of the century struck Europe and the United States. Blizzards halted travel throughout Europe. Half the birds of the Continent froze. The shamrocks froze in Ireland. Starving wolves left the hills in Italy and Yugoslavia to devour watchdogs in city streets. Spain lost her citrus crop, and twenty inches of snow fell in subtropical Sicily in the Mediterranean. In the United States cold and frost wiped out the vegetable crop in Florida and damaged 80 percent of the citrus crop. Ice floated on the Mississippi River as far south as Missouri. All the Great Lakes, even the largest, were virtually frozen over at the end of February.

As the sunspot number increased, weather remained strange. Commentators dubbed 1965–66 as "the winter the Ice Age returned to the Rockies," for snow was heavy in the American West. On Miami Beach a cold wave dropped the temperature to 2°C. and forced lifeguards to wear parkas. Blizzards brought the most snow in a century in Maryland, and closed offices in Washington, D.C. Helicopters were used to bring food to snow-stranded citizens in Syracuse, New York. In the American Midwest cattle froze or were suffocated when barns sheltering them were buried in snowdrifts.

In the summer of 1966 more than a hundred people died in a heat wave when a high-pressure zone settled over St. Louis, Missouri. Such blocks in the global wind system are common in a period of cooling world climate.

In January 1967 snow started falling over Chicago. It kept falling steadily until thirty-seven inches of the stuff buried the city. Transportation halted. Rail lines were cleared by jets of burning gas, and trains were used to haul snow from the city. Two days after the storm struck, another like it seemed headed from the Rockies toward Chicago, but it missed the city and dissipated. The Windy City nearly became the first U.S. city of the next Ice Age.

In 1968 and 1969 the sunspot cycle peaked. In the African Sahel a drought began that would last at least six years and claim 200,000 or more lives by famine. In 1965 nearby Ethiopia began to suffer drought and locust plagues. Weather in the U.S. and Europe improved.

Sunspots were minimal in 1971 and 1972, when bizarre snows in the Northern Hemisphere expanded the ice- and snow-covered area by 12 percent. During the minimum frightful rains flooded the Mississippi Basin and the Great Lakes, while drought began in the U.S. Midwest. Europe enjoyed warmer weather, but some floods devastated Italy. Lack of adequate snowfall hurt crop production in northern China, India, and the Soviet Union (which in 1972 purchased a large portion of the U.S. grain crop). By 1973 drought was affecting Great Britain, South Africa, Bangladesh, the Sahel, Ethiopia, several Caribbean nations, the Philippines, Japan, central Africa, northern Australia, and parts of the United States. Soviet crops failed in 1972, thrived in 1973, failed in 1974, and suffered their worst failure in a decade in 1975. U.S. harvests fared somewhat better.

In other words, the period 1962 to 1963 resembles 1971 to 1974 in terms of bizarre, anomalous weather, and the two periods are precisely one sunspot cycle apart. By extension we may assume that between 1981 and 1985—the next low point following a sunspot peak—weather will be either very cold or very dry or both. Likewise, 1976 to 1978 could mirror the snowstorms and heat waves and storminess of a decade earlier, and if this happens sunspots could be a part of the cause. Warming weather should prevail from 1979 to 1980, despite the gradual conjunction forming among planets in the solar system. By 1981–82 heavy snows could return to the Northern Hemisphere, as they did in 1971. Assuming the sunspot cycles are even and last eleven to twelve years, partial drought should visit the U.S. Midwest in 1985 to 1987, and severe drought should come in 1996 to 1999.

Of course, such predictions assume that sunspots alone determine our weather, which they do not. Whatever influence they have is modified by a thousand other natural and unnatural forces. But it makes good reading on a quiet evening to visit the local library and study newspapers of eleven, twenty-two, thirty-three, or more years ago to see if, where you live, odd weather follows a pattern corresponding to sunspot cycles.

The Changeable Sun

The sun provides our heat and light, so when climate changes on Earth a change in the sun is the most likely cause. Until recently the majority of weather scientists rejected this view, preferring to think sunlight was constant.

But studies of sunlight conducted during the past twenty-five years by astronomers at the Lowell Observatory in Arizona, and made public in July 1975, now clearly establish that sunlight intensity changes, and that the sun flickers like a candle in a breeze. The studies, begun in 1950, involved several scientists; since 1973 Wes Lockwood has been in charge. To minimize distortions in data caused by Earth's atmosphere, the Lowell research never looked at the sun directly; instead, measurements were made of sunlight reflected from the planets Uranus and Neptune and from Titan, the largest satellite of Saturn.

Since 1972, says Lockwood, all three of these bodies have brightened. The conclusion: sunlight has increased in the visible spectrum by 1 or 2 percent, or in the ultraviolet spectrum by a like amount. This cannot be accounted for by a change on the planets, because they brightened at the same time and in the same proportion. A change in sunlight is the simplest explanation.

Radiation from the sun increases by perhaps 2 percent at the peak of the sunspot cycle, according to Soviet meteorologists K. Ya. Kondratyev and G. A. Nikolsky, but this has been measured from Earth and may reflect changes in our ozone layer as well as changes on the sun. (A curious oversight in the U.S. space program was its failure prior to 1975 to launch satellites equipped to measure tiny changes in solar radiation.) Data is inadequate to prove them right or wrong.

As Lockwood noted, such changes might be subtle and come in parts of the spectrum—such as the ultraviolet range. In 1955 a Trans World Airlines meteorologist, Edwin Farthing, wrote that a shift in the ratio of red to green lines of ionized iron in the solar spectrum usually presaged by three days a cold front passing through the U.S. Midwest. This sunlight change, he said, coincided with the formation of atmospheric troughs in the North Pacific Ocean—a phenomenon like that described nineteen years later by Drs. Roberts and Olson.

In 1960 a Massachusetts Institute of Technology meteorologist, Hurd C. Willett, proposed that global climate changes were caused by changes in the ratio of three types of solar radiation —visible light, ultraviolet light, and corpuscular radiation, the charged particles of the solar wind. Earth avoids Ice Ages when ultraviolet light and corpuscular radiation are minimal, wrote Willett. When the proportion of ultraviolet light increases, the equatorial regions grow warmer, the poles remain cold, and U.S. weather gets wetter. Global cooling occurs when sunspot numbers are high and corpuscular radiation predominates. The poles then grow warmer in winter, which disrupts planetary wind patterns and causes severe storminess and extremes of temperature—and would cause an Ice Age if continued long enough. Willett wrote that sunspots vary in an eighty-year cycle, but that the decline of sunspot numbers after 1955 would usher in forty years of cooler, wetter weather in the United States.

Such sunspot analysis is as yet an occult art. Is Willett right? According to Dr. John A. Eddy of the National Center for Atmospheric Research, the coldest part of the last "Little Ice Age" in Europe occurred between 1645 and 1715, a seventy-year period in which astronomers reported virtually no sunspots at all. A sunspot increase marked the end of the recent warming trend. But why? Nobody is yet sure.

"If the sun's heat were to drop only 13 percent, the whole earth would soon be covered with a mile-thick mantle of ice," wrote astronomer Ernst J. Öpik in 1958. "Contrariwise, if the sun's radiation were 30 percent hotter than it is, life on the earth would be destroyed by a heat wave." Öpik, an Estonian working at the Armagh Observatory in Northern Ireland, assumed that changes in the sun shaped Earth's climate.

The sun, wrote Öpik in *Scientific American*, operates something like an atomic reactor. It is balanced when energy pouring from its surface equals that created by hydrogen fusion in its core.

But it has not been balanced, because at some point in the ancient past, metals began building up around the core. This slowed the escape of energy, created pressure, and ended when the sun's core burst outward with great force. The sun expanded greatly, but the result was less heat in the solar system because the spreading of the reactor core reduced the number of fusions per second in the sun.

Since that expansion, which probably happened near the dawn of time, the sun has been contracting, pulled into a tighter and tighter ball by its own gravity. As it gets smaller, it burns hotter, for the fusion process intensifies. When the sun was expanded, wrote Öpik, the Earth was everywhere frozen. But during the past 500 million years it has warmed as the sun has gotten hotter. As the sun keeps heating, he said, "the long-term trend on the earth should be toward a hotter and hotter climate" with Earth temperatures climbing .6°C. every 100 million years. Smaller changes, like Ice Ages, were probably caused by shorter-cycle "flickers" in the solar fire. A change of 8 percent solar energy output, which he suggested could be caused by metals concentrated near the sun's core, would be sufficient to cause an ice age or turn the Earth tropical and make New York as warm as Florida.

In the 1930s British physicist and Nobel Laureate Paul Dirac presented a theory that throughout the universe gravity was slowly decreasing. This led German scientist Hans Haber to an opposite conclusion from Öpik. If gravity was diminishing, then both Earth and the sun would be expanding as the power drawing matter toward their cores weakened. Earth would move farther from the sun as the force holding it in orbit weakened, and the sun would burn less brightly as it expanded. Thus Earth's climate was gradually cooling.

Haber assumes that Earth was once covered with dense clouds, like Venus today, and that diminishing sunlight eventually brought the clouds back to earth as water when cooling forced condensation of their vapor. His evidence is at times weird, but always interesting. To prove that life evolved under clouds he writes, "Even today jungle plants thrive best in the dim twilight of a tropical environment. They are the nearest modern relatives of the primitive plants of past eons."

Better evidence for a hotter sun comes from Mars, which according to the most recent photographs from Mariner space probes has dry river beds on its surface. Today virtually all water on Mars is frozen, but it was liquid when the rivers flowed. Was the sun hotter? Or did Mars have an atmosphere better able to hold what solar heat it received? Or has Mars changed its orbit, probably not to the extreme suggested by cataclysmic prophet Immanuel Velikovsky (who believes Mars has repeatedly moved near Earth's orbit and threatened us with collision) but to some degree? The simplest explanation is that solar output has changed slightly over millions of years.

Mars analyst William K. Hartmann of the Planetary Science Institute in Tucson, Arizona, assumes the sun has recently cooled and that it may run in warming-cooling cycles of perhaps 200 million years, varying in luminosity up to 15 percent during such periods. Two other Mars specialists, Drs. Joseph Burns and Martin Harwit of Cornell University, speculate that a tiny increase in warmth would melt frost on the red planet and set the rivers flowing again. This could be accomplished, they say, by human intervention to shift the orbits of Mars' moons, or by a few reflectors in space near the planet, or by nuclear power plants on the Martian surface—and this change could make Mars a habitable human colony able to support millions of Earthlings, complete with plants and animals.

What makes Dirac's analysis most credible is that in our universe gravity *is* weakening. Evidence for this comes from Thomas Van Flandern of the U.S. Naval Observatory, who in 1974 announced that the moon is moving away from the Earth at a speed of four centimeters per year. This, he told the American Geophysical Union, "is the first numerical result which appears to have as its most probable explanation that gravity is decreasing."

If this is true throughout the universe, then gravity is decreasing by about one part in 10 billion per year, and a person who weighs 70 kilograms (154 pounds) weighs about seven-millionths of a gram less each year. Each lunar month, the time it takes the moon to orbit Earth, grows longer—and the total grows by a two-thousandth of a second per year.

The changes seem tiny, but multiplied by billions of years they could account for vast changes in the relationship between Earth and the sun, which cools as gravity weakens. Work is now underway to compare Van Flandern's measurements with changes in the orbits of other planets, and with changes in light coming from several stars.

An even more revolutionary question baffles scientists: has our sun already burned out? The light radiated from the sun's surface was produced by fusion deep in its core 10 million years ago. Thus, the sun's nuclear reactor could have shut down thousands of years ago without our knowing it.

This notion was prompted by evidence produced by chemist Raymond Davis, Jr., of the Brookhaven National Laboratory, who for years has operated a huge chemical tank a mile underground in the Homestake Gold Mine in Lead, South Dakota. Davis's tank is designed to measure neutrinos, subatomic particles which carry energy and momentum. Scientists are unsure of what they are, exactly, but matter in the universe is transparent to them. Created by nuclear reactions in the sun's core, they shoot away, carrying about 5 percent of the sun's total reactive energy with them. Zipping through space at near the speed of light, they pass through the Earth as light passes through a window. The neutrinos that pass through the 100,000 gallons of cleaning fluid in Davis's tank—buried far underground where few other atomic particles are likely to reach—cause chemical reactions as they go by, producing Argon-37 and other new isotopes. By measuring chemical changes in the tank, Davis can "count" the neutrinos that have arrived.

Just as neutrinos can pass through Earth, so too do they pass through the sun. A neutrino arrives in Davis's tank only minutes after its creation in the sun's core. Heat generated by the same atomic solar reactions will float slowly to the sun's surface, be converted into light and other radiation, and not arrive on Earth for 10 million years. Thus, the neutrinos are messengers able to show what is happening inside the sun right now.

Calculations of how fusion reactions inside the sun work give precise prediction of how many neutrinos Davis should be able to "count," but thusfar he has never found more than 20 percent of the number nuclear physics calculations say *must* be there. Where are the missing neutrinos, and what does their disappearance mean?

Some scientists argue that neutrinos are little understood as yet, and could be lost or decayed by other atomic interactions. The simpler explanation is one feared by scientists: the sun's thermonuclear furnace has shut down, and neutrino production has virtually stopped. If this has happened, writes *Science News*, "the long-run result will be highly detrimental to the earth. Nevertheless, it seems unlikely that anybody alive now need worry about the consequences." This assumes the sun would cool thousands of years hence—but if all the energy pouring outward from the sun's core suddenly stopped, gravity would collapse the whole sun into its core in 17 minutes, according to physicist William Fowler of the California Institute of Technology.

What has happened? Dr. Fowler, who has studied neutrinos since 1959, proposes an explanation similar to Öpik's. The core of the sun goes through periods of heating, which causes expansion of the core, which then causes cooling. "The cooling would be so marked that it would practically stop the nuclear reactions," he writes. "The sun's luminosity would decrease. After a long time, the core would shrink, nuclear reactions would begin again, and the sun would regain its original brightness. The decrease and recovery of luminosity would take 10 million years, the same time required for energy to reach the sun's surface. Over most of this period, neutrino emission from the sun should be well below that observed by Davis."

Dr. Fowler surmised that this might explain the Great Ice Epochs on our planet, each of which has lasted for roughly 10 million years.

These epochs come at intervals, surprisingly regular, of about 200 million years. Perhaps, guessed anthropologist Robert Ardrey in his 1961 book *African Genesis*, this has something to do with a point our solar system crosses during the 200-plus million year revolution around the center of our Milky Way galaxy. In 1975 astronomer W. H. McCrea of the University of Sussex, England, turned the idea into formal theory. At certain points, he wrote, we pass through the spiral arms of the galaxy, and along these arms are giant lanes of intergalactic dust. The dust falls into the surface of our sun and provides extra fuel for burning. Thus, for a time the sun burns hotter. This increases precipitation on Earth and, in accord with the 1934 theory of Sir George Simpson (who also expected a hotter sun to create new Ice Ages), fuels massive snowfall on Earth and chills the planet. In 1939 British astronomers Fred Hoyle (now Sir Fred) and R. A. Lyttleton, and before them Harlow Shapley, proposed that the same thing would happen from our solar system passing through an intergalactic cloud of dust or hydrogen gas.

Core samples taken on the moon by Apollo XV astronauts give support to McCrea's hypothesis; they show that thick layers of micrometeoritic dust have arrived on the lunar surface at tidy but unmeasurable intervals. But is McCrea's reasoning correct?

As with all theories, this can be disputed. If dust in Earth's atmosphere dims our sunlight like dirt on a window, would not dust between us and the sun have a similar effect? What we accept often depends on what we want to believe. Optimists will like Dr. McCrea's analysis, which says we have just passed through the dust lane of the Orion nebula and that our climate should be warmer, because the sun will be slightly cooler, for at least 200 million years. In this instance, however, *warmer* and *cooler* are relative terms.

How shall we sort out all these bewildering, seemingly contradictory facts and theories? The scientists are still debating whether our climate will go on cooling, or start warming, or stabilize soon. By way of summary, we can assume that a combination of factors favor continued cooling in the immediate future:

1. Temperatures have been dropping in the northern half of our planet for the past 30 years. This constitutes a change in climate, according to standards set by the World Meteorological Organization. Once begun, such changes tend to build momentum through natural feedback mechanisms (e.g., increased albedo in the case of cooling) and hence are likely to continue.

2. The warm climate of the first half of our century was "highly abnormal," the National Academy of Sciences reported in 1975, being typical of climate during only 5 percent of the past 700,000 years. In some regions like the Gulf of Mexico, notes Brown University climatologist John Imbrie, temperatures have reached this century's warmth during only 2 percent of the last 450,000 years. Thus, a gambler would set the long-term odds at between 20 and 50 to 1 against our remaining as warm as we were between 1900 and 1950. Chances are that our planet will swing back toward a more normal, colder climate.

3. Climatic changes have followed clear cycles for millions of

years. As the National Academy of Sciences warned in 1975, if these cycle patterns continue as they have in the past, we can expect climatic cooling for the next 10,000 years.

4. The warming sunlight reaching us is changing in accord with subtle changes in Earth's tilt, wobble, and orbit relative to the sun. These changes now favor cooler summers, warmer winters—a regime of low seasonal contrast that favors more snowfall and colder climate.

5. Sunlight reaching Earth's surface is measurably diminishing, perhaps for both atmospheric and extraterrestrial reasons. This loss of energy in our global thermal budget can only lead to a colder climate. The solar energy that we know has already been lost to our planet by dimmed sunlight is many times the total human production of energy in recent decades.

Thus, further cooling is likely. Only a fool would gamble it will not continue, for what is being wagered is our future on this planet. The next question: how will it affect us?

PART II
The Human
Side of
Climate

5

Climate's Children

Between 1770 and 1820 the world turned upside down. Britain's American colonies declared and won their independence. In France revolution overthrew the king and removed other heads of government. And Eskimos landed in Scotland.

As in most other great changes in human history, climate influenced events. These years were among the coldest of the last "Little Ice Age," and the chill they sent across Europe compelled Great Britain to import more food, which required more revenue, which meant higher taxes on her colonies. But the harsh climate of her North American colonies, especially in rebellious New England, had helped breed toughness and independence in the people there. They resisted the tax. Britain sent troops. When war came, General George Washington's rebel troops suffered through several horrible winters, like that at Valley Forge, but such weather hampered British action and favored the Americans. During this period, winters were so cold that New York Harbor repeatedly froze. Heavy cannon could be dragged across the thick ice from Staten Island to Brooklyn. In 1780 every harbor in the North Atlantic coast was frozen shut during the winter.

In France the weather swung wildly, as it does in periods of climatic change like ours today. The late spring and early summer of 1788 were extremely hot, and in northern fields the grain shriveled. On July 13, with farmers ready to harvest what remained, heavy hailstorms struck and caused more loss. This led to food shortages, a prime ingredient in political turmoil then as now. The shortages begat the Paris bread riots of 1789 (and the response attributed to Marie Antoinette: the people want bread? "Let them eat cake."). This led to the storming of the Bastille.

The Eskimos must have been surprised to discover Scotland. They were on a fishing expedition, paddling their kayaks along the margin of the Arctic ice pack. But during these cold years the ice was much extended. Sea ice clogged waters south of the Faroe Islands, only 250 miles north of Great Britain, and blocked shipping lanes between Iceland and Greenland. Gulf Stream waters reach Europe by flowing near this region, and the winds that make much of Europe's weather blow across it. Scotland then must have seemed more like home to Eskimos than it would today. Nevertheless, the fishermen landed only long enough to look around and be seen by surprised Scots.

A century earlier, in the 1690s, northern Scotland suffered cold that destroyed harvests in eight consecutive years. Famine followed, and as many died from starvation as from the Black Plague. "The consequent weakening of the Scottish nation," writes climate analyst James D. Hays, "may have made union with England inevitable."

During this same period of the last Little Ice Age, England too was cold. The River Thames froze repeatedly, and during many weeks of the 1683–84 winter Charles II and his court walked from one side of London to the other across its ice.

How strange it seemed. Before 1400 weather had been unusually warm in England, and in Europe. French merchants complained about the competition from English wine, the product of vast English vineyards. But then came a cooling. In 1431, the year Joan of Arc burned at the stake, the Little Ice Age began with a winter that froze every river in Germany and devastated French vineyards. In the Alps glaciers began growing, and by the seventeenth century were engulfing villages in France and Switzerland. Some years, even some decades, between 1430 and 1850 were warm. In England the decade of the 1560s was exceptionally wet and rainy, while that of the 1660s brought drought that nearly dried up the Thames and helped kindle the Great Fire of London in 1666. By 1607 England was establishing colonies in the New World, where the climate seemed warmer.

The cooling had long before helped destroy another European colony in North America. Between 800 and 1400 the Northern

Hemisphere enjoyed warm climate, much as it did between 1880 and 1940. The warmth improved harvests and spawned a population explosion in Scandinavia, whose people started "Viking," their Norse verb for exploring, pirating, conquering new lands. The Vikings sailed up rivers into Russia as far as the Caspian Sea. They led raids against cities in the Mediterranean. And they established farming communities in several lush, fertile lands which they discovered, including Greenland—named for its bright grasses and warm soil—and Vinland—named for its wild grape vines. Vinland, we now know, was Newfoundland, where Viking artifacts dating from around 1000 A.D.—five centuries before Columbus landed in the New World—have been found.

How long the Vinland colony endured is unknown. Like the Greenland colony, founded in 980, it was established at the climatic optimum of these few warm centuries. Soon afterward sea ice and fog began increasing in the North Atlantic, and Viking ships found sailing routes to the new lands more and more difficult. Records of the colonies indicate hostilities with native Indians in Vinland and with Eskimos in Greenland. In both places the last Norsemen may have died by arrow or spear. But the climate change was what destroyed these colonies. Communications with Vinland were cut off by huge packs of ice, probably within a century after its founding. Greenland fared slightly better, but excavation of Norse graveyards shows a pattern: graves were dug shallower as the cooling intensified, apparently because digging got harder as the ground turned to permafrost. By 1492, the year Columbus began his first westward voyage, Pope Alexander VI complained that no bishop had visited Norse Catholics in Greenland for eighty years because sea ice clogged the routes. The last colonists there had died in 1450, and Greenland was covered with snow and advancing glaciers. The day of Viking raids against European coasts, like that of the Mongol and Moslem hordes riding into European lands from the East, ended in the cold.

The Vikings did not discover the New World. When they arrived it was already heavily peopled by Indians and Eskimos, whose ancestors came to the Americas during the last Ice Age. An Ice Age grows when the precipitation that otherwise would flow into rivers and thereby return to the oceans, freezes. Thus, as ice increases on the land, the level of water in the oceans falls. During the Ice Age that peaked 23,000 years ago sea levels dropped enough for a huge bridge of land to emerge between Alaska and Siberia. For several thousand years people, animals, and even plants migrated eastward across the relatively ice-free bridge and began populating the new continents they found.

When the Ice Age waned the snows melted and their waters poured back into the oceans. The Bering land bridge was submerged in the rising waters, and the stuff of legends emerged.

Was there a Noah's flood like the deluge spoken of in writings of the Hebrews, Babylonians, and the American Indians? Was there an ancient island kingdom called Atlantis (Aztlan by the Aztecs) that sank in one night from flood and earthquake? In 1975 University of Miami paleoclimatologist Cesare Emiliani suggested that these old stories may have been based in fact.

From his study of sea-floor sediments, Dr. Emiliani has found evidence of a rapid influx of fresh water that entered the Gulf of Mexico around 9600 B.C., precisely the date the Greek philosopher Plato used for the sinking of Atlantis.

What caused this flood? Emiliani believes that at one point during the retreat of the last Ice Age a brisk re-cooling occurred, sending a thin ice sheet surging across the Great Lakes and far into the United States. This sheet melted as it moved, flooding the Mississippi Basin, pouring vast amounts of water into the Gulf of Mexico and thereby raising sea levels throughout the world.

Much of humankind lived by the sea, as is true today. But then the sea level was 150 feet lower than today, and many settlements would have been destroyed. "We postulate that ensuing flooding of low-lying coastal areas, many of which were inhabited by man, gave rise to the deluge stories," wrote Emiliani and his colleagues.

After the Ice Age waned, advanced civilizations grew up among the North American mound builders, and later among tribes of the Pacific Northwest and Atlantic Northeast. On the high plateaus of Mexico the Aztecs established a formidable empire, as did the seafaring Mayas in Yucatan and Central America.

But climate remained a problem. While a warming world helped the Vikings claim new lands, the new climate brought severe drought to Indians in the American Southwest, who abandoned many once-irrigated pueblo cities to escape it, and to the Mayans, whose highly advanced civilization may have been destroyed in large part by droughts like those plaguing Honduras today.

And climate continued to shape culture. In the Americas emerged the *only* great equatorial civilization in history, the Inca Indians. But they should not properly be thought of as equatorial, for their empire was built many thousands of feet above sea level in the Andes Mountains, where climate more nearly resembled Northern Europe than the tropics.

We confront a good historical question: why do civilizations develop in the temperate zones, where winters are cold and summers warm, rather than in the heart of the tropics, where winter never comes? Nobel Laureate economist Gunnar Myrdal speculated thus:

> Almost all underdeveloped countries are situated in the tropical or subtropical zones. It is a fact that all successful industrialization in modern times has taken place in the temperate zones. This cannot be entirely an accident of history but must have to do with some special handicaps, directly or indirectly related to climate. . . . extremes of heat and humidity in most underdeveloped countries contribute to a deterioration of soil and many kinds of material goods; bear a partial responsibility for the low productivity of certain crops, forests, and animals, and not only cause discomfort to workers but also impair their health and decrease the participation in, and duration of, work and its efficiency. . . . *climatic conditions often impose serious obstacles to development.* [Myrdal's emphasis]

The equatorial regions have never given birth to an artist recognized as great in civilizations of the temperate zones, nor, as Aldous Huxley noted, to a poet of the stature of a Wordsworth. Why?

Two answers seem obvious. One, noted by Myrdal, is the power of the tropics to destroy the works of humankind. If an equatorial artist ever set brush to canvas, his paintings have long since rotted. Wood carvings and writings on paper would have met a similar fate, destroyed by insects and mildew and fungus—all of which abound in tropical lands where no annual winter suppresses them. Significantly north of the equator, Mayan artists carved in stone—a laborious task in humid tropical climes—and raised buildings, but nature has drastically eroded even their most recent artwork.

The other answer is that, just as the tropics abolish all human attempts to leave marks that will last through time, they also make the need for human inventiveness small. The tropics are a Garden of Eden, after a fashion. Vegetation is lush, and for natives content to live in jungle forests food is abundant. Systematic, organized work creating social edifices is as unnecessary as it is useless in such surroundings. Nothing lasting can be built, but nothing needs to be built. Thus, individuals cannot accomplish things, but merely *are;* the tribe in which identity is found exists beyond history and beyond plans for building the future; all reality is in the here and now. Mystics might deem such equatorial people wiser than those in "civilizations," but to people in industrialized nations the equatorial mentality is anything but civilized.

Inventor R. Buckminster Fuller has a theory of human evolution: that the need to condense our reality has expanded our minds. The first division among humans, says Fuller, came when some people decided to put everything they needed into a boat, challenge a new "fluid" dimension, and sail to new lands. Doing this involved compressing what they needed into a small, floatable form. Another division came when people started to fly, and reality was distilled into a form small enough for an airplane to lift. Most recently, humankind has been learning to fit its needs into space capsules for travel to the moon and planets. The people who will dwell in space stations a century or less hence, suggests Fuller, are the primitives of a new species of cyborg beings wholly merged with their tools, wholly supported by their own machines.

Whatever its flaws, this logic makes some sense when applied to climate. Our ancestors went separate ways, some staying in the tropics, some migrating into colder climates. Those in the cold learned to use caves, and later to build buildings, against the winter snows. Shelter was unnecessary in the tropics. Those in colder lands developed agriculture and irrigation to deal with erratic weather. This was unnecessary in the tropics, where either jungle or drought claimed all land, and where soil was too thin to cultivate for prolonged periods anyway (the Mayas and Thais tried and failed).

Thus, while it seems a paradox, people in the tropics were

totally at nature's mercy, although they lived by her grace. People nearer the poles, feeling constantly threatened by nature's whims, stretched their imaginations to win security from those whims; in the process they created civilization, a composite of values and tools mostly developed to make survival possible in harsh environments. In the cold, through many small achievements, humankind came to believe that it could "claim dominion over nature," and what we call industrialization is but the latest manifestation of this belief.

Climate shaped human values, human ideas, human actions. Thus, we are all climate's children, for she made us what we are. (The anthropologist-philosopher Loren Eiseley suggests that this climatic influence runs deep, that at the deepest level of the subconscious mind of all those descended from ice age people there swirls the genetic memory of an unending snowstorm.)

But dominion has never fully been achieved by any nation. Most of the early agricultural civilizations developed alongside rivers, quickly expanded their populations to the limit of food production, and collapsed when small changes in climate reduced water in the river or soil fertility became exhausted by the overplanting of one crop. Some theories hold that climatic change destroyed the early civilizations of the Hittites, Mycenae, Mesopotamians, and the peoples of the Indus Valley in India. That such changes could have contributed to their downfall is possible but unproven.

In 1915 the American philosopher Ellsworth Huntington wrote that civilizations tend to rise and fall in accord with a 600-year climatic cycle. The reason: global redistribution of good and bad climate causes "nomadic eruptions," like the Viking eruption of 1000 years ago. His theory has been widely debated. Some of the migrations he analyzed in Chinese history and attributed to climate changes, for example, were more likely caused by plagues or by the introduction of new crops like yams and peanuts that allowed the expansion of agriculture to new lands. But he has also acquired many imitators. The most recent, Nels Winkless III and Iben Browning, suggested in 1975 that "nomadic eruptions" had been caused by climatic cycles of 800 years, beginning in 3600 B.C. Owing to lack of information, such theories are neither provable nor disprovable.

Empires have not risen and fallen randomly, observed Huntington. The center of empire, at least in the West, has moved north —from Egypt to Greece and Rome, then to France and England, and, with allowance for settlement of the Americas by northern Europeans, may now be moving to Moscow. Why? In 1924 arctic explorer Vilhjalmur Stefansson wrote a book advancing Huntington's observation and embellishing the idea with his own native Canadian chauvinism. "Civilization," said Stefansson, "has coincided in part at least with [humankind's] march northward over the Earth into a cooler, cleaner, more bracing air." Equally apt perhaps, empire has to some degree coincided with cool-temperate climate. At its height of power Egypt was temperate, not desert. The Soviet Union gained its power during the brief climate warming of the early twentieth century. Or perhaps Buckminster Fuller was right.

Even today most nations depend on rainfall for their food growing, either directly or as runoff in rivers. A five-year drought could turn even the United States, the world's biggest food exporter, into a food-importing nation, and a change in climate could cause such a drought. The drought in the African Sahel lasted from 1968 until 1974, and rainfall there has not yet returned to previous levels; the Sahel drought, say experts, was a product of global climatic cooling.

In the ancient world Egypt endured climatic change better than all other civilizations. The reason: the Nile River, whose yearly floods watered and provided new soil for the land, came north from the tropics. During the height of the last Ice Age the Nile was reduced to a trickling stream because drought struck equatorial lands. But after the Ice Age the world warmed to a climatic optimum in about 6000 B.C., and from then until the building of the Aswan Dam by Egypt in the twentieth century the Nile floods were dependable and let the nation grow abundant food.

The greatest military achievement of a young Roman Empire was the conquest of Egypt in the first century B.C. Egypt then was the granary of the Mediterranean, and by owning it Rome was assured of food. Thus the Caesars could take men from farming and put them into armies, and thus they could conquer the world. The Roman Empire eventually expanded to boundaries set by climate, to the line of severe frost and winter snow in the north and to deserts in the south. Climate at this time was much like ours today, but the lands of North Africa were greener and better able to grow crops. From about 400 A.D. until the next climatic optimum neared in 800 A.D., climate throughout Europe cooled somewhat but remained mild.

A thousand years after the final collapse of the Roman Empire, cooling climate prompted several European nations to try building empires of their own. This new imperialism has been ascribed to mercantilism, greed, and the desire to explore and exploit new lands. Just as accurately it can be ascribed to the severe cold climate that had settled over Europe, causing famines on a regular basis and forcing abandonment of many northern villages on the continent.

The European quest for empire was a quest for a transclimatic base, for lands where food and raw materials could be grown, or where revenue could be earned with which food could be purchased. Its prime targets were tropical and subtropical lands in Africa, Asia, and the New World—areas nearer the equator than Europe, and hence warmer.

The more such lands a European nation controlled, the more flexible its options became, so the desire for new territory was fierce, as was the competition. When British colonists rebelled in America, France—defeated by the British in their bid to control Canada and the Great Lakes—provided military aid to the rebels and ultimately guaranteed their independence.

Imperial politics, based largely on analysis of climate and territory and resources, became exceedingly subtle. A case in point: the United States Civil War. London, which lay farther north than the entire United States, originally established its subtropical Southern American colonies to produce key raw materials, including cotton. After American independence, Great Britain continued to buy Southern cotton; her mill industries depended on it.

In 1850 the South nearly declared its independence from the North in America, but war was forestalled by a series of political compromises. British policymakers recognized that a civil war in the new American nation would come soon, nevertheless, and this would endanger the cotton supply. Britain immediately moved to plant the entire Nile Valley in northern Egypt, which her empire controlled, with cotton. When civil war erupted in the United States in 1860, British mills had an alternative cotton supply.

Such are the ironies of history. Had the Confederacy declared its independence and gone to war in 1850 it would have won. Britain, to protect her only supply of vital cotton, would have entered the conflict on the side of the Confederacy, and the British fleet would have blockaded and shelled Northern ports. But because she had time and other subtropical territory, Britain was able to diversify her sources of supply and avoid fighting a war. (Acquiring such territories, of course, meant she had to fight other wars.)

The young United States was busy in the nineteenth century building her own internal empire in the New World, buying land from France and Russia and seizing land from Mexico. The country was different from what Americans today would recognize—cooler and wetter in most regions.

The year 1816 was the coldest year in North America's past two centuries. Snow stayed on the ground in New England and upstate New York all twelve months, and farmers called it "the year without summer." The year before, a volcano had erupted in Iceland, and its dust and smoke high in the atmosphere dimmed sunlight. But from that year until 1940 the Northern Hemisphere slowly warmed.

American pioneers moving west had a common complaint: across the Great Plains the prairie grass was so high that one covered wagon could easily lose sight of another in a wagon train and get lost. Bison thrived on this tall grass, called buffalo grass. But as the climate of the continent warmed, rainfall decreased, and the buffalo grass slowly disappeared. This, say some climate experts, likely had as much to do with the near-extinction of immense bison herds in America as did the butchery by Anglo hunters.

The new land was warming and drying, and by the 1930s much of the land where the buffalo grass grew was a dust bowl, and winds sent dust aloft in huge swirling clouds over the fragile plains and over the valley of the Mississippi, the Nile of the New World. The planet was entering a new stage of climatic history and a new collection of empires was emerging: one in North America, one in the long-frozen lands of Russia, and one, fated to be short-lived, in the warming country called Germany.

The warming climate lifted them from the ice, and soon may cast them down again. How soon? In the 1940s global climate began cooling. . . .

6

The Seeds of Our Own Destruction

Between 1880 and 1950, Earth's climate was the warmest it has been in five thousand years. In that brief space we created a new, fair-weather world.

Harvests improved everywhere, and farmers moved north to cultivate virgin lands once too cold for agriculture. Food grew abundant and, feeding upon it, world population nearly tripled. Wars and revolutions tore old empires apart. New social systems claimed rulership over most of humankind.

It was a time of optimism and wrong assumptions. We assumed that our climate was warm and unchanging, that therefore we could always grow enough food, and that the supply of cheap energy was unlimited.

The optimism has shriveled in the first chill of the cooling. Since the 1940s winters have become subtly longer, rains less dependable, storms more frequent throughout the world. Food production has suffered. Oil companies have been forced to divert crude oil from gasoline production into home heating oil. Fuel and food prices have been rising sharply, weakening the economies of most nations.

The seeds of these present problems were sown before the cooling began. Many decades ago experts warned that population growth would outrun food production, and that oil would become costly when people realized how little of it our planet possessed. But these seeds grow faster in cold weather. The cooling shortens the time we have to develop solutions, and that may make our problems unsolvable. Relatively few people will freeze to death if the cooling trend begun in 1950 continues until 2050—but perhaps two billion humans will starve to death, or will die of the symptoms of chronic malnutrition.

"The world's food-producing system is much more sensitive to the weather variable than it was even five years ago," said Dr. James D. McQuigg of the National Oceanic and Atmospheric Administration's Center for Climatic and Environmental Assessment in 1975. According to McQuigg, the generally good harvests and food production increases of the past 15 years came not so much from improved grains and technologies as from extremely favorable temperatures and rainfall. "The probability of getting another fifteen consecutive years that good is about one in ten thousand," he says. According to NOAA climatologist J. Murray Mitchell, Jr., "From the agricultural productivity point of view, the climate's not going to get better. It can only get worse. . . . If there's anything we can be reasonably confident about in terms of projections of future climate, it is that the climate of our crop-growing areas will become more variable than it has been in the recent past." Weather will worsen if the cooling continues, and crop losses will increase.

This must be emphasized: the cooling's threat to humankind is not confined to the distant future, when glaciers will march halfway to the equator. It is immediate, measurable in lost bushels of wheat and corn and in barrels of oil diverted from other purposes to heat homes and factories and offices. In the time it takes you to read this book, at least a thousand people will have starved to death *because of* the impact climatic instability already has on food production. But by some indications the cooling has scarcely begun.

Some would call this statement extreme. Even in 1974, a year when bad weather devastated crops in many parts of the world, an equal division of Earth's harvests (before losses to pests and rot) would have provided 666 pounds of grain for every person on the planet, an adequate diet of nearly two pounds daily. If all of the world's potentially arable land produced food as efficiently as farmland in Iowa does today, wrote the president of the American Association for the Advancement of Science, Dr. Roger Revelle, in 1974, Earth could support 60 billion people, fifteen times the number alive today. Why worry?

The answer is complex and involves a number of factors—most of which are influenced by climate. It is neither optimistic nor pessimistic, but realistic. It is not intended as doomsaying, for the problems can be dealt with if humankind accepts certain costs and risks. Here we will consider what will happen if today's problems continue, for this is what must be weighed in the balance when those costs and risks are judged.

The Overfed, the Undernourished

The average American and Russian each eat 2000 pounds of grain each year, a fact that will surprise many Americans and Russians. Most is eaten indirectly, in the form of meat. To produce a pound of beefsteak, a cow is fed between 7 and 13 pounds of grain; to produce a pound of meat protein, a cow eats 21 pounds of protein in grain, yeast, fish meal, or other nutrients. In the United States today livestock eat six times more protein than Americans do, and as much grain as all the people in India and China eat.

The average human being in Africa or Asia eats 400 pounds of grain yearly, almost all directly as grain. In consequence, the United Nations Food and Agriculture Organization (FAO) estimated in 1974 that at least 460 million people, one human in eight on the planet, were "chronically malnourished." What does this mean? Lack of food has put nearly half a billion people into a twilight zone of survival, able to reproduce but too poorly fed to produce or progress. As British science writer Jon Tinker said in 1974, chronic malnutrition "means that you go blind from Vitamin A deficiency; or that lack of iron makes you anemic and breathless; or that you suffer from rickets or beri beri, scurvy, goiter, or pellagra. It means that your children grow up stunted, not just physically but mentally as well: often listless, vacant, and slow on the uptake. It means that malaria, or cholera, or even a bad cold can kill you. Chronic malnutrition means the end of hope. It reduces human beings to the level of animals, ready to snarl over a crust of bread, locked in a cycle of breeding and death."

Since 1970 food riots have erupted in India, Pakistan, and several African nations. In the United States, an official of the Los Angeles Police Department said in 1974 that his forces were training for food riots, in case transportation faltered and the four-day food supply in urban markets was taken by hoarders. The fear of food shortages now extends from the richest nation to the poorest.

Food is not equally distributed on this planet, nor is it likely to be. Quite the opposite, the disparity between overfed and undernourished nations will increase. In 1961 world food reserves could have fed humankind for 95 days. By 1974, following massive crop losses caused by bad weather, world food reserves fell below 24 days. Food surpluses are shrinking. Food production costs are increasing. Developed nations now hoard food or use it as a political tool in bargaining with allies and enemies, or sell it to improve their international balance-of-payments position relative to rising oil prices. The amount of food given to poorer nations in aid is rapidly declining, and the global climate change is making it harder for them to grow food of their own.

Two Babies per Second

"The human race did not reach its first billion people until A.D. 1830," wrote Harvard professor of nutrition Jean Mayer in 1975. "Only 100 years later, in 1930, the second billion was reached; 30 years later, in 1960, the third; and it took only 15 years, until this year, for mankind to add the fourth. This accelerating, alarming increase is due to a drop in the death rate."

Why did the death rate decrease? In part because of modern medicines, but population has expanded in regions where such medicine has scarcely been applied. Another compelling reason is the increase in global food supplies, the fruits of a gradual warming in world climate that began in 1817. The changing climate warmed lands in the high latitudes and, by influencing global wind patterns, improved rainfall in many lands nearer to the equator. Excellent climate combined with improved agricultural technologies after World War II to expand food supplies even more, and population increased to meet available food.

In the 1970s the global cooling that began three decades earlier has begun to hurt food production. But population growth has its own momentum and continues.

Demographers debate how fast population will increase, with numbers that numb the imagination. Optimists say that Earth will have 6 billion people by year 2000—three for every two alive today—and 8 billion by year 2050. Pessimists say that 8 billion people will be sharing the planet by the turn of the century, less than twenty-five years from now, and that 16 billion or more —four for every one alive today—will inhabit Earth by year 2050. Optimists say that birth control programs can stabilize Earth's population at between 6 and 12 billion. Pessimists say that the population growth will continue, its rate accelerating, until humankind plunges over the brink of worldwide famine or war.

At present growth rates, estimates biochemist and author Isaac Asimov, Earth's population will be 15.2 billion by year 2044; 20 billion by 2059. "Our planet will contain all the people that an industrialized world may be able to support by about 2060 A.D.," he says; "That is not a pleasant outlook for only 85 years from now."

Rapid population growth is no more equitably distributed than food. Even as a baby boom swelled populations in the developed nations following World War II, it accounted for but 20 percent of the growth in world population. Today population growth in the United States is near zero. In West Germany and much of the Soviet Union, population is declining. More than 90 percent of all future population growth will occur in the world's poorer countries. By 1985 Bangladesh will grow from 84 million people to 124 million, unless famine, plague, or war change present trends. By 1985 India will gain an additional 207 million people, nearly the number in the whole United States.

Optimists used to say that even huge population increases were tolerable because each new body came equipped with hands to hoe soil and produce food. But much of future population growth will be urban, for the less developed nations are experiencing an exodus from the countryside to the city. City dwellers do not grow food; they must buy it, beg for it, steal it, or have it provided by the government.

The twelve fastest-growing cities of the world are in underdeveloped nations. All will more than double their populations in the fifteen years leading to 1985. By that year Bangkok, Thailand, will have more than 7 million inhabitants; Karachi, Pakistan, more than 9 million; Seoul, South Korea, more than 10 million. Buenos Aires and Rio de Janiero will each be home to nearly 11 million;

Bombay, India, to 12 million. São Paulo, Brazil, will have nearly 17 million people; Mexico City nearly 18 million.

Even without war or enforced birth control, say optimists, such population growth can be controlled. Stable populations in industrialized nations prove that prosperity can remedy the problem. Well-fed, educated, secure people tend to have few children.

Even if this were true, say pessimists, it means nothing. Relative to the developed world, the standard of living in less developed nations is falling. Improving it will be nearly impossible. Industrialized nations have never faced what now confronts these poor countries: population that will double within four decades. When population doubles, food supplies must double just to keep a nation in the same place in which it started. When city population doubles, urban services such as water, sewage, welfare, police, electricity, schools, and other necessities must double just to maintain the previous standard of living. Additional energy must be used to bring food from the countryside. In recent decades many poor nations have had huge percentage increases in food production wiped out by population growth. Running as fast as they can, these nations fall farther behind.

Even world prosperity could mean disaster, say some pessimists. America's success involves control of between 10 and 30 percent of the world's total developed resources by 5 percent of its people. Thus, by simple arithmetic we find that to bring the whole world to the American level might require a 200 to 600 percent increase in resources, which would strip the Earth bare if we could achieve it at all. Moreover, the average American generates fifteen times more pollution and garbage than a fellow human in India. The environment could scarcely stand the strain of world pollution at the American level.

Nothing consumes like success. Those families that have found prosperity in poor nations may or may not have fewer children. But almost invariably their eating habits change; they give up grains and roots and start eating meat—which means that they eat at least seven times more grain than when they were poor. (And where Western values have invaded, even the poor have changed habits; in many parts of Africa, for example, mothers now deem breast-feeding undignified and buy formula for their infants, thereby wasting a precious natural resource and spending needed money for an unnecessary, costly substitute.) One price of curbing population is more food, even if we see this optimistically.

These problems aside, evidence suggests that a rise in the living standard in poorer nations produces "*rising*, not falling, birth rates," at least during the transition between peasant and industrial economies. Thus concluded an American Association for the Advancement of Science report issued in 1975. Such societies value large families; children provide old-age security and joy for parents. Even with prosperity, this value will take several generations to change. Parents, meanwhile, will celebrate their newfound prosperity by raising the biggest families they can. The result will resemble America's postwar baby booms. If these findings of his research team prove correct, said panel chairman Dr. Peter F. M. McLoughlin, then population will inevitably outrun food supplies, and "somebody is going to have to die."

At present more than two children are born per second, and world population grows by at least 6 million per month, 72 million per year. Of this increase, 90 percent is in the poorer nations where the cultural response to thousands of years of high death rate has been big families. For these nations, yesterday's salvation may be tomorrow's doom. If the global cooling trend continues, most will be hit by alternating floods and droughts. The changing climate will reduce their populations.

The Land That Is No More

Food can be grown anywhere, even on top of Mt. Everest. The only problem, says Worldwatch Institute President Lester R. Brown, is the price. Such Himalayan wheat might cost $10,000 a bushel.

Only 7 percent of Earth's total land mass has the right natural topography, texture, nutrients, and climate—temperature and rainfall—to permit normal agriculture, according to agronomist William Paddock. To grow food elsewhere requires what ecologists call "energy subsidies," human-supplied inputs such as fertilizer, irrigation water, soil conditioners, land terracing, and artificial heat, to make up for natural deficiencies. A wheat field atop Mt. Everest would need all these subsidies, plus protection from wind and solar radiation.

With such subsidies, more land becomes usable for crops. But how much more? By the 1969 estimate of Michigan State University food production specialist Dr. Georg Borgstrom, 30 percent of the Earth's land can be cultivated, being within climatic limits of arability. Another 30 percent is "productive," but unsuited to agriculture—forests, poor pastures in Australia, the African bush, steep slopes and mountainous regions. By comparison, in 1975 Philip Abelson, editor of *Science*, estimated that "for the world as a whole 45 percent of the land is suitable for crops but only 26 percent is used."

These figures are not necessarily contradictory; they reflect two views of what is a realistic energy subsidy. If we were willing to pay the price, we could grow food everywhere. Recall two of the above figures: 30 percent of Earth's land is potentially arable without extreme subsidies, and 26 percent is already under cultivation. Humankind already grows crops almost everywhere it easily can. The cost of farming marginal areas will be high. The cost for the worst 40 percent of our land, the deserts and "frozen deserts" near the poles, would be immense.

Untilled lands, moreover, are poorly distributed for feeding the world's population. According to Abelson, in Asia and the Far East 84 percent of all usable land is already in crops, but only 23 percent of such land is now farmed in Latin America. Thus, "there is little room for expansion of the farming area in Bangladesh and India but a large potential in this hemisphere."

The United Nations Food and Agriculture Organization (FAO) wants Earth's cultivated land increased by 20 percent by 1985. This goal will necessitate growing crops on soil now considered marginal. Tropical forests will be cleared to bring new land under the plow. These thin soils evolved in ecological balance with forests; they depended on trees for nutrients. With forests gone, tropical rains quickly leach minerals and other vital elements from such soil. Much tropical soil is lateritic; it turns rock-hard when plowed up and dried. It makes good building material (the famous Angkor Wat temples were built partly of Cambodian laterite, the product of an extinct empire's unwise cultivation that exposed fragile native soil to open air), but it is poor stuff for farming.

In Java all the easily farmed land is being cultivated, but population keeps growing at a rapid rate. Farmers have started plowing hillsides, mountainsides, and newly cleared jungle lands. Massive erosion is the result. Streams and dams have silted up, so that every heavy rain brings flooding. The effect is a net loss of food production. Under Premier Nikita Khrushchev the Soviet Union during the 1950s and early 1960s invested more than 10 billion dollars in a "Virgin Lands" plan to make marginal areas in Siberia produce crops. The results: vast areas were devastated by drought and became dust bowls, and repeated crop failures despite the huge investments of energy and money was a factor in Khrushchev's 1964 ouster. Some experts believe that cultivation of most other marginal lands on Earth will have equally bad consequences.

Some marginal lands could become productive, but only with expensive technologies—irrigation, chemical fertilizers and pesticides, soil treatments, new seedgrains, and modern machinery. An imitation of such American methods might cost a small Bangladesh farmer $2000 per hectare* or more, scarcely a worthwhile investment where government controls grain prices and water distribution. For much less, perhaps $80 per hectare according to experts, he could greatly increase crop yields. But even this lower figure offers little hope, for in 1974 the average Southeast Asian farmer could afford only $6 per hectare. FAO says it could meet its goal with an investment of $9 billion per year between 1974 and 1985. This is possible in theory, but unlikely as a practical matter of politics. The few nations that could pay the bill have little incentive to do so.

Seventy-one percent of Earth's surface is water, and many see hope in harvesting the seas. Since 1945 the global fish catch has tripled, and it probably can be sustained at twice the present level—but this is still a tiny fraction of land-grown food.

Salvation will not be found in aquaculture. During the 1970s climate-caused changes in ocean currents have slashed fish harvests—hectoring sardine fishing by Norway and California, shrinking the anchoveta harvest off Peru's coast, and precipitating a "cod war" between Iceland and Great Britain.

The oceans, far from being vast expanses of food, more nearly resemble huge deserts with a few fertile oases. Oceanic food growing needs sunlight and nutrients, just as on land. But in the

*A hectare = approx. 2.47 acres.

oceans nutrients are concentrated in deep layers of water near the bottom, and sunlight sinks no deeper than the top surface layers. Rich harvests abound only where deep water upwells into the sunlight. With few exceptions, this happens only along certain seacoasts where ocean currents churn against land, such as the Grand Bank east of North America or the zone of the El Niño current off Peru. Ships sail halfway around the world to fish in these few rich areas.

The emerging global scramble for food explains why many nations now declare that their boundaries extend at least 200 miles out to sea. Seafood thrives in these coastal waters. In many instances the declarations are in response to greedy invasions by powerful nations, such as the Soviet Union and Japan, whose fishing fleets use the most modern technologies to suck up whole schools of fish. But changing climate has prompted the new limits, too. Cod thrived in the ocean currents around Iceland until a few years ago, for example; then the currents shifted more than 100 miles south. Iceland expanded her sea boundary to follow the currents, even as nations might change a mutual boundary if the river between them changed course. Ecuador and Peru have done likewise, especially since 1972 to 1973 when the global climate changed the temperature of the El Niño upwelling and cut the anchoveta harvest in half.

Sea harvests can be increased by using seafoods not widely eaten now, such as squid and trashfish, and by applying new methods. This gives little hope to hungry nations, for they lack the capital to harvest such crops. The rich nations, able to afford new fishing technologies, already consume roughly half the global sea catch, and their share will increase. The Soviets, for example, now have trawlers in the Antarctic harvesting planktonlike krill. In the U.S.S.R. it is processed into a protein-rich paste. Acceptance by Soviet consumers is poor; like much American-caught seafood, it is relegated to cattle feed, to supply the growing Soviet taste for meat.

But the limit of naturally produced seafood is in sight. To take more than double the present catch of most species, experts agree, would cripple their ability to reproduce. Even this estimate may be too liberal; the powerful climate-induced cooling in the North Pacific and Atlantic oceans is already slowing fish and plankton reproduction and reducing numbers. But in all likelihood nations like Japan and the U.S.S.R., having invested heavily in their new fishing fleets, will keep catching as they can. In consequence, wrote Jon Tinker in 1974, "catches will . . . triple over the next decade—and fall disastrously after that."

The areas we harvest must be thought of in time as well as space: how many weeks of the year can food be grown? The global cooling has already shortened the growing season in Great Britain and parts of Canada by at least two weeks. Longer winters and shorter summers in Iceland have cut hay yields by at least 25 percent. In the Northern Hemisphere in 1971-72 snow has expanded to an area 12 percent larger in winter than in 1968. As a result, regions that once grew crops are now barren or give poor harvests; some regions that once gave two good crops a year now yield one at best. Storms and unreliable rain hurt crop production in most parts of the world. In the African Sahel and Ethiopia drought has already brought famine as cooling global climate pushes monsoon rains toward the equator.

The changing climate is already reducing the amount of arable land on Earth, measured in hectare-days, and making more and more land "marginal" for agriculture.

But even without the cooling, and with all the new lands tilled by FAO efforts, cultivated land will remain roughly the same while population increases drastically. More mouths must be fed from each bit of land. In 1974 three members of the Club of Rome, Professor Mihajlo Mesarovic of Case Western Reserve University, Professor Eduard Pestel of the Technical University of Hanover, Germany, and French government economist Maurice Guernier, published their computer study of what this will mean for the next fifty years.

North America, they estimate, will remain fairly stable: in 1975 every square kilometer of its cultivated land fed 105 people, and by year 2025 will need to feed only 175 people. In Western Europe the increase is sharper: the same land that fed 320 people in 1975 must support 485 people fifty years hence. These are the lands of stable population, able to feed themselves using existing technology in the near future.

Prospects are worse elsewhere. In Latin America the square kilometer that fed 195 people in 1975 must support 575 people by

2025. The land in China that fed 750 people must by 2025 feed 1180. In South Asia, where the greatest increase is expected, the square kilometer that fed 370 people in 1975 must feed 700 by year 2000, and by 2025 feed an incredible 1,370—a 370 percent increase in fifty years!

Mesarovic, Pestel and Guernier compute that population will outrun per-kilometer yield in South Asia. Between 1975 and 2025 the daily protein supply is expected to drop from 64 grams per person to 37 grams. For many, this loss in nutrition will be fatal. In 1980, they expect, an increase in child mortality will begin rising quickly; this will peak in about 2010, when in one year 40 million South Asian children will die, nearly half from starvation or chronic malnutrition. Ten million children died in this region in 1975, estimate Mesarovic, Pestel and Guernier, but only 1 percent of them died from lack of food. Many were undernourished, however, for the minimum daily protein ration they assume an adult needs for good health is 70 grams.

The Seeds of Hope

Between 1960 and 1971 world food production grew by 2.8 percent per year while world population grew by only 2.4 percent. The food increase was not enough to overcome widespread malnutrition; by FAO estimates, it must grow by 3.8 percent every year for the next several decades to do that. But it was enough to prevent mass famine and feed optimism.

In 1972 optimism withered. Unusual weather, the consequence of changing global climate, caused crop failures in many parts of the world. Storm and drought struck in the Soviet Union, flood and drought in the United States, India, Africa and Australia. In 1973 summer snow fell on Canadian wheatfields. At least 400,000 people starved in Ethiopia and the small nations south of the Sahara Desert. Global food reserves, which in the past had insured against such catastrophes, fell by more than two-thirds, and nations with surpluses started using them for politics and profit. Food aid usually given to poorer nations was cut back.

Until 1972 optimists held a dream called the "Green Revolution." The world could feed itself, they believed, if only every bit of arable land could become as productive as the average farm in

Iowa. New hybrid grain, fertilizer, irrigation, pesticides, and technology could make it happen. Through 1971 the Green Revolution appeared to be winning the race with hunger, and even such nations as India were approaching self-sufficiency in food production. Since then a host of counter-revolutionary forces have crippled it.

The new hybrid grains, developed by agronomists such as Nobel Laureate Dr. Norman Borlaug, could produce much more food per unit of land than could old grains under good conditions—abundant water and chemical fertilizer. During the 1960s rainfall was generally adequate in most parts of the world. New wells tapped virgin water tables, allowing irrigation for the first time in some parts of India and Bangladesh. Fertilizer was cheap. Crop yields increased, in the optimists' word, miraculously.

But if these hybrids were better than older grains in good times, they were worse when conditions changed. They lacked the toughness bred into natural grains by thousands of years of evolution. When rain patterns began oscillating between flood and drought in the 1970s, yields plummeted. With weather changes, too, came temperature changes, and as a University of Wisconsin research team found, the new hybrids were more sensitive to such changes than were old-fashioned grains.

The hybrids were grown from only a few genetic "parents" picked for high yield and no other qualities. Nature, warn ecologists like Dr. Paul Ehrlich, a Stanford University biologist, relies on genetic diversity for its resiliance; if a virus emerged that preyed on one genetically weak hybrid, and if half the world was planted with that hybrid, the consequences would be devastating. "The Irish potato famine of the last century is perhaps the best-known example of the collapse of a simple agricultural ecosystem," Ehrlich and Dr. John Holdren wrote in *American Scientist* in 1974. "The heavy reliance of the Irish population on a single, highly productive crop led to 1.5 million deaths when the potato monoculture fell victim to a fungus." Thanks to the Green Revolution, such a collapse could happen today on a global scale. The cooling makes it probable, for a chilling climate destabilizes balanced ecological systems—killing some predators, favoring some victims, encouraging the survival of genetic mutations. This is one reason plagues have often in the past accompanied periods of climatic cooling. But we have gone ahead and replaced whole native populations of seedgrains with new hybrids that have never stood the test of time.

A glimpse of the possible future consequences of such policies appeared in the Philippines during the 1960s. Farmers planted their fields with a new Green Revolution high-yield rice strain called IR-8. Almost immediately the crop was wiped out by a disease called tungro. The farmers replanted with another "miracle" hybrid, IR-20, which soon succumbed to grassy stunt virus and brown hopper insects. They replanted a third time, with a hybrid called IR-26, which endured all the diseases and pests of the tropical land—but was destroyed by the strong winds. "When farmers clear a field of primitive grain varieties," said United Nations Environmental Program geneticist Reuben Olembo, "they throw away the key to our future. Eventually the new hybrid becomes so pure it cannot sustain itself."

Ecologists such as Ehrlich and Holdren believe that modern agriculture is at odds with nature. Nature mixes many species together in field and forest, where they support one another in balanced ecosystems. Humankind, trying to maximize productivity rather than natural diversity, clears land to plant only one species, then uses vast amounts of energy killing "weeds" and "pests"—other plants and animals—and providing energy subsidies such as water and fertilizer to make up for missing nutrients that a natural system would have provided. The human tendency away from natural diversity makes crops weaker and hence more vulnerable to mass devastation. It often strengthens pests and kills friends; our insecticides, in addition to poisoning the human environment, have created stronger varieties of plant-eating insects but have in the past decade reduced the world population of honeybees, which help pollinate many crops, by 12 percent. It means we must pay higher and higher costs in energy to grow the same amount of food, for we must keep increasing our energy subsidies.

Iowa's success is the world ideal, but its productivity is built largely on luck and oil. One year's harvest on one Iowa hectare requires at least 200 gallons of petroleum—in the chemical fertilizer that feeds the crops, in the pesticide that protects them, in the tractors that till and the mechanical harvesters that reap. Even so, Iowa is lucky, blessed with naturally rich soil and usually adequate precipitation. During dry years some Iowa farmers irrigate, which means more oil to pump water to the plants.

This is "capital-intensive" agriculture, built on huge investments of money.rather than manpower. Without expensive artificial inputs to the growing process, the "miracle" grains would grow poorly—more poorly than natural grains. Nevertheless, in 1971 Iowa methods seemed exportable to the less developed nations. By late 1973 they no longer were.

The Arab oil embargo dealt a near-fatal blow to the Green Revolution. Oil was the prime source of chemical fertilizer, on which the revolution depended. When oil prices jumped more than fourfold, nations scarcely able to afford fertilizer in 1971 found themselves out of the market by 1974. Among the hardest hit nations was Bangladesh, which relied on Japan for fertilizer. Japan's first response to the embargo was to stop fertilizer shipments to Bangladesh (although for political reasons she kept up shipments to China). This was doubly sad because, after a few years of use, the food returns from chemical fertilizers diminish. A ton of fertilizer applied to land in Bangladesh would grow twice the food it would on long-fertilized land in Japan.

This problem has hit other developed nations, where inputs that yielded big crop increases a decade ago are now showing diminishing returns. Many fields have reached a fertilizer ceiling; to use more would cause severe nitrogen poisoning of the environment. In the United States, for example, the Iowa farmer during the years 1946 to 1970 increased his corn yield by 240 percent; but during the same period, to improve yields, he had to increase energy inputs to the land by 310 percent. This research, done by Dr. David Pimentel at Cornell University, suggests the difficulties that technological food production faces; even without more mouths to feed, we must invest ever-increasing amounts of increasingly costly energy to get the same amount of food.

Another type of fertilizer is available, made from mined phosphate. As fate would have it, this limited natural resource is concentrated in the Moslem countries of North Africa. Sixty percent of the world supply has been controlled by Morocco, which in 1975 negotiated a takeover of Spain's last African colony, Spanish Sahara—which controlled 20 percent of the world supply. Even before it gained an 80 percent control over world supplies, Morocco was able to force phosphate prices up from $14 a ton to $68 a ton since 1973. Today phosphate is a monopoly for all intents and purposes, subject to the same political (and religious) pressures and prejudices as Arab oil.

Pesticides are also based on petroleum, but even where available they are facing insects more resistant to old poisons. Using U.S. Department of Agriculture figures, Dr. Pimentel found that 34 percent of America's farm crops have been lost in the field to insect, pathogen, and weed pests in the 1970s; another 9 percent have been lost in storage to insects, rodents, and micro-organisms. In total, 43 percent of America's crops are lost before humans have a chance to eat them. Poorer nations routinely lose more than 40 percent of their crops to pests in the field. In some tropical countries, where storage is difficult, total crop losses often run above 50 percent. In the tropics no winter cold comes to kill insects, and pests proliferate; frequent rain makes conventional insecticides more difficult to use in safe amounts.

If global climate keeps cooling, the demand for petroleum will increase—nations near the poles wanting it for heating and to offset diminishing crop yields; nations nearer the equator needing more than 100 gallons per hectare to improve crop-growing capacity. This demand will push prices higher and higher, guaranteeing that poor nations of the world will be unable to afford fertilizer or pesticide and unable to save the capital needed for industrial development that would let them buy food on the world market. The International Monetary Fund in August 1975 announced that the trade deficit of the poorest nations increased more than 400 percent between 1973 and 1975, to $35 billion. This means they will be in beggary to the oil and food producers for the foreseeable future.

"If major drought occurs in the world now, or a crop failure in any of the large grain-producing areas of the world, tens of millions could die in an international disaster," wrote Dr. Borlaug in 1974, "and little could be done to prevent it."

In 1800 the British Rev. Thomas Malthus predicted such a disaster. He theorized that human population increases faster than food production, so that a sizable portion of humankind lives always on the brink of starvation. A small change in climate or food or oil prices may be enough to push millions over that brink.

A number of stopgaps have postponed his prophecy. In past centuries people could migrate to new, warmer lands; this option now is closed. In recent decades we have planted new grains, dug new wells, used new fertilizers and pesticides. But the grains are increasingly vulnerable. The wells tapped water tables that in the absence of dependable rain will soon run dry. The costly fertilizers are reaching the limit of their effectiveness. The pesticides are creating stronger pests. The optimists are praying for another miracle to again forestall the Malthusian calculus.

We have sown the seed of vast population, and the vulnerable seed of new grains. These may prove to be the seeds of our own destruction. Hope persists. More technological tricks are on the drawing boards, some of which might work, none of which have the massive financial backing that is needed to develop them quickly. The nations with the most money to spend are those least likely to suffer in the next several decades of unstable climate. In building our fair-weather world, we have expanded population far beyond what natural food sources can provide for. Our prospects for global well-being in the emerging foul-weather world of the cooling are better than hopeless but far less than bright. Two billion starved by A.D. 2050? This could prove to be an optimistic estimate.

7

Climatocracy:
The Politics
of Global Cooling

Climate is power, because it is a vital natural resource like oil, coal, iron ore, or uranium.

If a nation has a good climate, it can cheaply grow enough food for its people. Water is abundant. The land is warm enough in winter and cool enough in summer for human comfort. Vast amounts of costly energy need not be spent heating and cooling buildings.

If a nation has bad climate, food is hard to produce. Rain is undependable, water scarce, growing seasons short, and destructive storms commonplace. Food must be imported at high cost. Precious energy must be spent irrigating land, warming and cooling buildings, clearing snow from highways and railways to make commerce possible.

As much as any other single factor, climate determines a nation's economic condition.

Since its birth in 1945, at the dawn of the global cooling, the United Nations has gained more than 140 members. Of these only 10—the United States, Canada, Australia, New Zealand, Argentina, South Africa, France, Rumania, Burma, and Thailand —have been able generally to produce more food than they consumed. All the rest, in years of bad weather, have been forced to beg or buy food from these few self-sufficient nations. If the cooling continues, bad-weather years will come frequently. World food production will drop, even as world population grows. The wealth of nations will flow to those with food to sell. The cost of food will go up like a moon rocket, and only those who can pay will eat.

The cooling promises for Earth's future an extreme form of what we had in the past: climatocracy—the rule of nations blessed with climatic resources over nations poor in such now-dwindling resources. For the next fifty years or so, this means a world dominated by the bread-and-butter imperium of the United States, whose plowshares will have the power of swords.

The United States and Canada between them already supply 85 percent of the world's internationally traded grain. Thus North America holds more of a global monopoly on food than the Organization of Petroleum Exporting Countries (OPEC) does on oil. The U.S. produces more than three times as much food as Canada now, and if the cooling keeps moving downward from the North Pole it will destroy Canadian crops far faster than those in the United States. Food could become virtually a U.S. monopoly.

And how does the United States use its food surpluses? As a "new international weapon," said Agriculture Secretary Earl Butz at the World Food Conference in November 1974. In effect, he was saying that the fruits of America's good climate would be used to political advantage.

"Food *is* a weapon and we should use it," said Daniel P. Moynihan, former White House advisor and U.S. ambassador to India, in February 1975. "We would have been shocked to use it a decade ago, but now we should." If OPEC nations were willing to embargo oil as a political weapon, the U.S. was willing likewise to use food. Apparently he spoke in a way the government approved; four months later Dr. Moynihan was confirmed as ambassador to the United Nations.

The new weapon had been unsheathed during the World Food Conference. In November 1974 a drought-caused famine was sweeping India, and there were frenzied food riots. From the conference Secretary Butz and Senator Hubert Humphrey wired the White House, asking that as a humanitarian gesture the U.S. give India an additional million tons of grain. The president refused, but a week later he announced that grain would be given to Syria, an Arab nation bordering Israel. In congressional testimony later that month, University of Wisconsin geophysicist John S. Steinhart assailed the action, saying, "Even if the reasons for the choice are diplomatic, [the president has] decided that some in India will die and some in Syria will live."

The choice had political overtones. In a December 1974 article, *Science News* quoted options for food assistance just prepared by the U.S. State Department, the most favored of which would "enable us to meet the Egypt and Syria political requirements" and give small increases to Israel and Jordan, but would provide only a "minimum essential level of programming to Bangladesh, India, and Sri Lanka."

The United States had used food as a political weapon before, but never as declared policy. During the 1950s the Eisenhower Administration, for example, held up emergency grain shipments to India while bargaining for mining rights to her vast deposits of thorium, a radioactive cousin of uranium believed to be viable stuff for a new nuclear weapons technology. Pressure from public opinion restored the grain shipments, and the U.S. got no thorium rights.

The 1972 credit sale of a quarter of the total U.S. wheat crop to the Soviet Union as a part of *détente* was a political act with severe consequences: it raised food prices for Americans, and it sent world grain prices soaring beyond the reach of nations where people already were starving.

The U.S. refusal to sell grain to the Marxist government of Chilean President Salvador Allende in 1972 was a political act that, say analysts, "contributed to the turmoil in Santiago and eventually to his overthrow." And, following crop losses to excess rain, the 1972 decision to embargo soybean exports was a political act that brought hardship in many Asian nations where soybeans are a staple food; in the U.S. soybeans are fed to livestock.

Although political, these actions were impromptu, caused by immediate events. The new U.S. policy, by contrast, makes the political use of food a regular, all-inclusive tool of diplomacy. Less developed nations increasingly will find themselves squeezed between OPEC's "oil weapon" and America's "food weapon." They will have strong reason to find a weapon of their own.

He Will Live. You Will Die.

The embargo by Arab oil producers helped forge America's "food weapon" policy, but another rationale is quietly used among policymakers to justify it. This rationale is summed up in one little-known word that will be in common usage by 1985—*triage*.

Triage is a procedure used in battlefield hospitals. The incoming wounded are divided into three groups: those who will live, regardless of treatment; those likely to die, regardless of treatment; and those who can be saved by treatment. Those who can be saved then are treated. Those who cannot are left to die if resources are limited.

The word was first applied to world food and population problems by William and Paul Paddock in their book *Famine—1975!* and later by Dr. Paul Ehrlich in his best-selling *The Population Bomb*. All three see world population outgrowing food supplies by a wide margin. Thus many millions of people are going to starve, inevitably. Because the United States controls the bulk of exportable food, it will decide who lives, who dies, and whether fated famines happen soon or later. The later famines come, the more people will starve. America's task is to feed people where hope of eventual recovery and self-sufficiency exists. Where it does not, the natural force of famine must be allowed to reduce populations.

When these authors proposed triage in the late 1960s they were called fanatics. Within a decade, however, their ideas were gaining support. On the eve of the 1974 World Food Conference, the president of the National Academy of Sciences, Dr. Philip Handler, said, "Cruel as it may sound, if the developed nations do not intend the colossal all-out effort commensurate with [feeding the world], then it may be wiser to 'let nature take its course' as Aristotle described it: 'From time to time it is necessary that pestilence, famine and war prune the luxuriant growth of the human race.' " Questioned about whether or not he advocated triage, Handler replied, "That's what I was saying, gently. . . . But it ought to be a damn cold-turkey decision—we should not drift into it. We must be aware of the consequences." Given continued rising prices for petroleum, needed for the fertilizer and pesticides on which the "Green Revolution" depends, added Handler, "I can't imagine not doing it."

Reaction was strong to Handler's remarks. "Revolting!" said British economist Lady Barbara Ward Jackson. "Inhuman!" said Philip Abelson, editor of the American Association for the Advancement of Science journal *Science*. Both believe population problems can be brought under control without massive famines.

But others have begun to take the "unthinkable" idea of triage seriously as a necessary option for U.S. policy. During congressional hearings in November 1974, University of California ecologist Garrett Hardin said that if Western nations keep giving food to poor nations, "the extra food will be converted into more babies."

In 1968, in *The Population Bomb*, Ehrlich likened Earth's "uncontrolled multiplication of people" to cancer, the uncontrolled multiplication of body cells. Giving food, he said, treated the symptoms but worsened the problem. His prescription was to "cut out the cancer" by giving food only to countries that controlled their population growth. Hardin echoed this in 1974, calling population expansion a cancer and saying, "You can't cure a cancer by feeding it." He recommended using food aid as "bait" to prompt rigorous family planning in other nations.

Also in 1974, Dr. Wayne H. Davis, zoology professor at the University of Kentucky, told a congressional committee: "It is time we recognize the absurdity of today's world and stop pouring rat hole money into a hopeless population sink. You cannot solve a hunger problem by feeding hungry people."

The "carrying capacity" of Earth—the number of people available resources can support at a given time—may already be past the point of no return. "The carrying capacity of the world is probably less than the number of people already in existence," University of Wisconsin ecologist Grant Cottam told Washington lawmakers in 1974. "We managed to survive without serious famine only because of the presence of a large food reserve. . . . With no reserves left, there is no way that we can avoid massive famines. The Earth simply cannot continue to support an exponential population growth." Food supplies that year, estimated Cottam, fell short of needs equivalent to 130 million people.

Dr. Georg Borgstrom, a Michigan State University food scientist, told Congress in 1974 that a billion more people are expected on the planet by 1984. Unless eating habits are changed in the developed nations, the only way these people could be fed would be by cutting down the world's remaining forests to provide crop-growing land, he said. But doing this would increase the already-great danger of major ecological disaster.

Dr. John S. Steinhart told Congress in 1974 that in the United States ten calories of fuel are required to produce one calorie of food, hence the jump in food prices when OPEC raised petroleum prices. Food in 1972 took 15 to 20 percent of the average American's income, but by 1974 it took 30 percent or more, he said. And costs have continued to rise. Consumption of resources such as oil must level off or decline, for such things exist only in limited quantities on the planet. "Only madmen and some economists think otherwise," said Steinhart.

Thus, these experts made clear, the U.S. must cut its own consumption, and must limit exports, of food. It is too costly to give away, and too expensive to sell at prices the less developed nations can pay. The United States used to give away surpluses to keep grain prices high at home and save storage costs. By 1974 food aid to other nations was shrinking. "We already practice triage, whether or not we so name it," Steinhart told congressional lawmakers, few of whom were surprised.

To judge by what the United States says, its policies are anything but triage-oriented. At the World Food Conference, Agriculture Secretary Butz said America was willing to provide technological assistance to developing nations, to recognize the need for "prudent" global food reserves, to seek cooperation with the Soviet Union in providing crop estimates, and to help remedy immediate world food shortages.

In his speech before the conference, U.S. Secretary of State Henry Kissinger was more specific. He called for world food reserves of 60 million tons over current levels. He committed the United States to a tripling of its contribution for the existing eight international food research centers by 1980, plus an expansion of its own research and an effort to "mobilize America's talent" through a new intergovernmental Food and Nutrition Research Program. He also pledged $5 million to initiate an internationally coordinated program in applied nutritional research, and $10 million toward programs to eradicate nutritional diseases such as vitamin A blindness and iron-deficiency anemia. "Now our consciousness is global," promised Kissinger. Many applauded his statement. Many wondered at his meaning.

To judge by what the United States does, the promises seem hard to believe. At the conference nations agreed to establish internationally coordinated, nationally held food reserves, the details of which were left to producer countries. But the U.S. policy declared by Secretary Butz is that agricultural production can be increased only by having "reasonable" prices for crops; i.e., high prices for consumers. To insure this, Butz believes, reserves should be privately held and subject to free-market forces; otherwise, food stocks would be victimized by politically motivated speculation, the dumping of food supplies to depress prices to farmers.

In response, the director-general of the U.N. Food and Agriculture Organization (FAO), A. H. Boerma, told reporters, "any purist concept of international free trade in food is dangerously outdated." But of course U.S. policy is not for free trade, but for high profits *and* other political objectives. Wheat was denied to the Chilean government of Marxist Salvador Allende, notes science writer John Douglas, "even when it offered cash." Sales and aid were restored after Allende's overthrow. Likewise, U.S. foreign aid, including food, to India was cut off when she invaded East Pakistan, now Bangladesh. A pure free-trader would presumably sell to anyone anytime.

Bad weather and crop failures—the product of Earth's unstable climate—have combined with political manipulation to influence supply and demand. In recent years the export price of rice has more than tripled, that of wheat more than quadrupled. In 1971 the U.S. sold less than $8 billion worth of foodstuffs to foreign customers. By 1974 she was selling $22 billion worth of food, double the amount she earned by selling weapons. These profits, owing to reduced free aid and higher food prices, largely offset balance-of-payments losses caused by increased OPEC oil costs. As long as such balance is needed, American politicians will see high food prices as in their national interest. Moreover, as the national secretary of the National Farmers Union told a Senate subcommittee, "Many farmers view permanent scarcity of food as a goal that would be appropriate to their self-interest." If American farmers went on strike for two years, a billion humans would starve. The same number may starve over the next few decades if rising prices put food out of their reach.

What of the short-term help Secretary Butz promised for nations facing famine? At the conference, writes food analyst Frances Moore Lappé, "the United States was not only unmoved but seemed downright resentful of urgent calls for even a modest increase in our food aid. But the fact is that we could *double* our current food aid, and it would amount to less than the $2 billion or more we gained in 1974 as a result of the higher prices we received from foodstuffs sold to the developing countries."

American food aid abroad, she writes, fell in 1975 to scarcely one-sixth the amount of food given in 1965, despite large population increases in the poorest countries. The United States, which always prided itself on feeding the hungry, now ranks next to last among Western nations in the amount of aid given relative to gross national product. And even at that, only 20 percent of its $1 billion food aid budget goes to nations facing mass famine. Almost all goes to such nations as Syria, Egypt, Jordan, and, since its government changed, Chile. Before communists seized their governments, about half of all U.S. food aid went to South Vietnam and Cambodia.

Even these figures deal only with food aid authorized, not delivered. Corruption in many countries diverts grain into the criminal black market where if it reaches the poor at all it will cost them dearly. In 1975 U.S. officials revealed that many grain shipments to foreign nations had arrived short of grain, or with cheaper grains substituted for what was to have been sent, or with tons of dirt mixed in to increase sale weight. Criminal indictments were handed down in some of these cases.

What of Secretary Kissinger's promise of increased aid to research? In a February 1975 article titled "Whatever Happened to Food Research," *Science News* studied federal budget requests and found that food research funds were being *reduced* by 9.5 percent. When questioned, state department officials said they were "trying to get a new congressional mandate for research," and that an error had been discovered in their budget request for the international food research laboratories. The journal recomputed, using the new figures, and sardonically reported the results "would mean that the decrease in total state department funds for food research would be 4.6 percent."

The total U.S. food research budget was to increase by 11 percent; because this was roughly the rate of inflation, the real budget stayed unchanged. But as *Science News* noted, of the $367 million requested, only 35 percent was dedicated to increasing food production, and only 9 percent was for international food subjects.

In January 1975 the chairman of the National Academy of Sciences Board on Agriculture and Renewable Resources, Dr. SylvanWittwer, said crop yields from fertilizers were tapering off, and unless better methods of crop production and population control are soon found "we could be living on borrowed time." A "Manhattan Project" was needed, said Wittwer, to improve plant efficiency—a commitment comparable in funds and expertise to the development of the atomic bomb. "The future well-being of mankind could be at stake," he said.

Many such warnings have come from calm, quiet scientists, but no major commitment of the sort Secretary Kissinger promised has been undertaken. Quite the opposite: in 1975 Kissinger reduced his call for a commitment of 60 million tons of grain in world reserves to 30 million tons, thus cutting America's share of promised but unestablished reserves by half. Likewise in April 1974 Kissinger had told the United Nations General Assembly that

the poorest nations, already beset by manmade disasters, have been threatened by a natural one: the possibility of climatic changes in the monsoon belt and perhaps throughout the world. The United States proposes that the International Council of Scientific Unions and the World Meteorological Organization urgently investigate this problem and offer guidelines for immediate international action.

But no such guidelines were drawn, no investigation was made, despite 400,000 deaths in Northern Africa from prolonged drought. Short of a major commitment to deal with such problems, NAS President Philip Handler implied in 1974, U.S. policy inevitably becomes triage. Millions could starve during the next few decades. Short of such a commitment, there will be too little food for everybody. American food allocations, made for political and profit motives, will determine who among the hungry lives and dies.

Bombs and Bread

Aldous Huxley, the visionary author of *Brave New World*, pondered in 1946 what might happen if new technologies gave the U.S. and U.S.S.R. control over global food supplies. "That their monopolies of food surpluses will be used as weapons in the game of power politics seems more than probable," wrote Huxley. "It will also be a great temptation to exploit a natural monopoly in order to gain influence and finally control over hitherto independent countries, in which population has outstripped the food supply."

Three decades later, Huxley's expectation is near fulfillment. The United States produces roughly 67 percent of all food available for international exchange. Its closest competitor (if that term applies), however, is Canada, which produces 18 percent of the world total. Until its own recent crop failures, the Soviet Union insisted that Eastern European allies rely on it for grain—even when it seized their harvests first—and this served as a bond of political dependence. But now the Eastern European nations are becoming dependent on U.S. and Canadian grain, and the United States is moving quickly to force a similar grain-dependence on the Soviet Union itself through "five year plans" of grain selling. A decade from now the Soviet Union's economy will be geared to require U.S. grain shipments, which can be cut off at any time.

Worsening climate and unstable weather have hurt Canada and the Soviet Union more than the United States, which is nearer the equator. Canadian wheatfields since 1970 have suffered unusual flooding and midsummer snow. Bizarre swings of weather in the Soviet Union, described by her scientists as the worst in more than a century, prompted huge purchases of U.S. wheat in 1972 and 1975. To some degree the Soviets can influence world grain prices with their purchases, but as world climate grows more variable the U.S. monopoly will become strong enough that it can dictate prices—at least for the next few decades.

A study of food as a strategic weapon done by the Central Intelligence Agency and made public in 1975 assumes that world grain shortages will increase in the near future. This "could give the United States a measure of power it never had before —possibly an economic and political dominance greater than that of the immediate post World War II years," when America had a monopoly on nuclear weapons and the absolute power to decide who lived and who died on the planet. The CIA's comparison is by no means far-fetched.

The CIA advised against use of the food weapon, warning that it would lead to retaliation in kind from nations controlling rare mineral resources. But this advice is being ignored. By 1975 the U.S. government had released all reserve "land bank" acreage for production to increase the American share of world food production. This release led to bountiful total harvests in 1975, the result in many cases of declining yields per hectare but much more land under the plow; such bounty cannot be expected to last for long. In January 1976 the United States officially declared that it would consider cutting off aid, including food aid, to any nation that voted against American interests in the United Nations.

The holder of any monopoly over a vital natural resource could be ruthless, attaching political strings to every deal and making puppets of all trading partners. Arab oil producers in the 1970s made some sales contingent on other nations' willingness to degrade diplomatic relations with Israel (e.g., Japan), and favored corporations that used advertising dollars to promote pro-Arab views (e.g., Gulf Oil).

Imagine the emergence of an OPEC-like cartel calling itself FOPEC, for Food Producing and Exporting Countries (or Companies). Like OPEC, it could sell to developed nations at high prices, then use the profits to purchase key industries in those nations, to buy political influence, and otherwise to improve its position.

With less-developed nations FOPEC might adopt a strategy of ransom designed to make them FOPEC colonies in a world system of bread-and-butter imperialism. During alternate decades the FOPEC cartel would supply large amounts of free or cheap food, provide life-sustaining medical aid, offer chemical fertilizers on easy credit, and thereby stimulate a large expansion in the population. Then it would cut off all such aid until the young, the old, and the sick in these nations began to perish. When the endangered governments begged for help, the cartel would be able to exact a maximum political and monetary price.

Any resemblance between these hypothetical strategies and U.S. policies in recent decades is purely coincidental. True, U.S. aid has done much to swell population in nations like India. It has helped destroy small, self-sufficient farmers in such countries, replacing them with expensive modern farming that needs machinery, replacement parts, costly fertilizers produced from oil, and other things the U.S. offers for a price. True, U.S. policy at present seems aimed at trading food for cash or political cooperation, and seems ready to withhold food from nations of whose policies it disapproves. But the thousands of humane Americans who have worked to prevent famine and malnutrition in poor nations never intended this to happen; they sincerely wanted to help others.

Help can sometimes hinder. Apart from the food and medicine that fed population explosions, what the U.S. has always offered poorer nations is its technology. But technology is more than machinery; it is a merger of tools and human values. "Become like us and you can feed yourselves" has been America's message and ideology. In other lands this gets mixed reactions. "Certainly it would be nice to have your soil and climate and food surpluses," one common response comes back, "but even if we could get those things, we have no desire to imitate your military-industrial complexity or materialistic values."

Inasmuch as American success is a product of doing things in the American way, however, the two are wedded and must be accepted or rejected as a unit. Imagine the cultural anguish in a newly independent nation facing this choice. Food can be had, but only if alien values and gods are embraced. Tribal societies must reform into competitive, profit-seeking patterns. Self-sufficient farmers must take jobs in cities, must pursue dollars although as children they lived by barter. Duty, routine, and time-clocks must replace more natural rhythms of sunrise, full moon, and the seasons. "Become like us . . ." The message meets with resistance, for it represents a kind of conquest, a surrender of cultural and national sovereignty.

Politics, as scholars define it, is nothing more or less than "the authoritative allocation of value." When a small nation embraces American technology, and hence American values, its politics have been taken over almost as surely as if U.S. Marines had landed. The banners in the capital may carry the emblems of multinational corporations rather than the American eagle, and the tribute exacted may be in corporate profits rather than formal taxation, but the effect is the same. The little nation becomes a puppet state, and any attempt to rebel earns retaliation.

Consider one such value: the common belief in developed nations that population growth is bad and should be restricted, especially in the less developed countries. Triage advocates, such as Dr. Garrett Hardin, suggest that the most humane thing the United States can do is use food as "bait" to induce family planning in other nations.

Obviously, large families have cultural approval in nations with fast-growing populations. For thousands of years, when early death claimed most children, the stork ran to keep ahead of the grim reaper. Even today, in such countries, children are viewed as good fortune, a source of added family income, a kind of social security for aging parents. Enforced family planning would be resisted by people with such values. Moreover, in lands with large numbers of Roman Catholics, artificial birth control methods would be resisted as sinful. And in lands with large numbers of Marxists, birth control would be resisted, by Chairman Mao Tse-tung's teaching, as an imperialist weapon used to hold down rising peoples.

Those who advocate using food as bait to coerce birth control readily admit that it would be a dictatorship, of sorts. But they add, pragmatically, that compared to mass starvation it is benevolent and humane.

How might such a food-baiting program work? In 1969 Dr. Ehrlich proposed that nations start propagandizing against large families, changing tax laws to discourage reproduction, and instituting birth control instruction in schools. They should increase funds for population regulation and environmental science, and reduce funds for "short-sighted programs on death control." The goal: lower birth rates, and higher death rates. American aid could be made contingent on a nation doing such things.

If such measures proved insufficient, Ehrlich suggested "the addition of a temporary sterilant to staple food, or to the water supply." Governments would control access to the antidote for this mass-ingested sterilant, and would issue only as much as needed to achieve the desired population. The antidote, he said, could be distributed by lottery.

If such a sterilant were perfected, would a benevolent global food dictatorship require its widespread use in India as a condition of food sales or aid? Perhaps. In countries where other population control efforts have failed, chemical sterilization could be rationalized as the best chance for national well-being.

But any such policy has frightening implications. Political elites would be tempted to use control of the antidote to increase their power. Political enemies could be denied parenthood; under the polite guise of licensing, individuals could be declared "unfit parents" before they had children. Black markets for antidotes would emerge, and officials could demand bribes for restoring a natural right. Those who took the antidote and then failed to become parents, or who bore deformed children, would blame the chemicals. Any enforced policy to limit population will seem oppressive, discriminatory, and to some genocidal. Many would resent such a policy, especially if imposed directly or indirectly by a foreign power. Some would resist violently.

Two tendencies must be noted in overpopulated nations: they are politically unstable, and they can no longer be regarded as powerless. India provides an example many others soon will follow. In 1974 she detonated an atomic device. To build it, her government diverted vast resources that might have gone into food production; political leaders decided that the atomic bomb was a better investment in India's future. In 1975, following years of food riots and threatened with mass protests demanding her resignation, Prime Minister Indira Gandhi—the mother of India's A-bomb—declared martial law, jailed political opponents, and claimed absolute political authority in the name of a national emergency. Critics called her actions insane; supporters said she prevented chaos, which in a nation verging on famine might have meant the deaths of millions. At what point, Bertrand Russell once asked, will a person trade the right to vote for a loaf of bread? Democracy is the first casualty when famine threatens.

Why would a small nation want atomic weapons? To be able to

bargain more nearly as an equal with the world superpowers: if the left hand holds an atom bomb, the begging bowl in the right is more persuasive. To be more secure: the standoff between the U.S. and U.S.S.R. that emerged during the 1970s, combined with the defeat of American aims in Indo-China, makes treaty guarantees from the superpowers less than comforting. To buy something precious at low cost: as with any evolving technology, the cost of atomic arms has dropped to the bargain-basement level. A small country can cheaply get the prestige and power that go with being the first or second in its corner of the world with A-bombs.

Misconceptions still surround atomic weapons: that they require secret technologies, that only rich nations have radioactive materials, that their cost is prohibitive, and that A-bombs are worthless without missiles or bomber aircraft. In reality, all necessary atomic secrets can be bought for less than $20 from the U.S. Government Printing Office. All "peaceful" atomic reactors, used for research or to generate electricity, create by-product plutonium that can easily be diverted and converted into atomic bombs.

The U.S., Canadian, and West German governments, among others, have been spreading reactors around the world at a reckless pace, spurred by profits that will top $50 billion by 1980. At least forty-nine nations now have reactors operating or under construction. In June 1975 Libya announced that the Soviet Union was selling her a large reactor. Days earlier, Brazil announced a deal with West Germany that included reactors, uranium-enrichment facilities to allow Brazil to process her own newly discovered uranium deposits, and a plant to extract plutonium from spent reactor fuel. The deal caused concern in Brazil's neighbor, Argentina, one of whose diplomats said, "We wouldn't dream of building a nuclear bomb—unless Brazil does." In March 1975 roughly 50 kilograms of plutonium—enough to make nine atomic bombs—disappeared from an Argentine atomic power plant; in August the International Atomic Energy Agency, which prides itself on accurate policing of reactors, dismissed the loss as a bookkeeping error. Informed sources report that the fissionable material used in India's first atomic blast came from a reactor purchased from Canada, but it could as well have come from the twin Tarapur reactors built by General Electric corporation.

America's crash Manhattan Project to create the atomic bomb

cost $5 billion, but costs have dropped since then. One expert study conducted in 1968 indicated that a poor nation with an atomic reactor could set up processing and weapons-fabrication laboratories able to make not just one, but a continuous flow of A-bombs, for as little as $125,000. The typical controlled U.S. price for plutonium in the past decade has been $10,000 per kilogram. A crude atomic bomb requires a bit more than five kilograms (11 pounds) of the stuff, $50,000 worth. Any nation on Earth could afford such prices, as could many corporations and individuals.

Governments, of course, do not sell plutonium to individuals. But as *Science* observed in April 1971, a global illegal market in plutonium is taking shape. Even at the controlled price, plutonium is five times more valuable than heroin, ten times more precious than gold. The incentive to steal is tremendous. Some nations and individuals might pay $1 million per kilogram or more for it. Add to that another million dollars to buy the services of one of the thousands of scientists in the world able to make a workable crude device, and the cost of an atomic bomb at black market prices is still scarcely one-tenth of one percent that of America's first A-bomb. Astute international criminals may soon start selling prefabricated atomic bombs to all who will pay the price. Oil-rich Libya, for example, has said she wants to have atomic weapons. In 1975 Libyan troops seized territory from neighboring Chad to gain several uranium mines.

Sophisticated delivery systems for such weapons are unnecessary. Because they may weigh less than 1000 pounds, the bombs can be carried to intended targets by terrorists in small private boats, airplanes, or even Volkswagen vans. Former Deputy Director of the Pentagon's Defense Atomic Support Agency Dr. Theodore Taylor estimates that one such device—even if a "fizzle" exploding with only one-tenth kiloton force, as compared to the 20-kiloton blast at Nagasaki during World War II —could kill 125,000 people if it were set off at the right hour in front of New York City's World Trade Center. Because raw plutonium is the most cancer-causing substance known, such a crude bomb would do more long-term damage than a relatively "clean" military weapon; it would spread throughout Manhattan a lethal radioactive pollution with a half-life of 24,000 years.

The technology of gunpowder made feudal societies built around hilltop castles obsolete. When it reached the hands of American colonial rebels two centuries ago, it started the decline of the British Empire. Likewise, the dawn of atomic techno-terrorism will make present concepts of national defense and international power obsolete. Missiles, anti-missiles, and vast armies give no protection against terrorist criminals armed with mega-weapons. Wars will consist in one nation secretly giving nuclear weapons to terrorists in another nation. In 1965, for example, the People's Republic of China offered two crude atom bombs to Al-Fatah Arab terrorists in Syria; fortunately, the Central Intelligence Agency intercepted the offer, threatened China with direct retaliation, and thwarted the deal. Future offers to terrorists coming from organized criminal groups might go undiscovered or be unstoppable.

What this has to do with Earth's changing climate is suggested by Table 7-1, which lists countries that will suffer severe cold or drought, either of which would hamper food production. Since the 1973 oil embargo, most nations have sought atomic reactors as an alternate source of needed energy. But these nations especially, as world population grows and shrinking food supplies prompt price rises, will have strong incentives to improve their international bargaining position. Atomic weapons offer a quick, cheap means to this end. The countries marked are those best able to acquire nuclear materials within their own borders. All the nations listed could potentially buy such materials from international criminals.

Other nations, not listed, could be indirectly affected by a changing climate. Japan, like Great Britain, is an industrial island nation poor in natural resources. She must buy more than 60 percent of her food with the products of her industry, but to keep her industry running she must import more than 99 percent of her oil. As the cooling continues, her food production will fall. Global demand for heating oil and petroleum-based fertilizer will greatly escalate prices.

Squeezed between demands of OPEC and FOPEC, Japan may decide to fight for her independence. She has done so before. Her 1941 attack against the U.S. fleet at Pearl Harbor was in response to an American-led oil embargo against her, joined by Great Britain (which controlled the Middle East), China (with some oil

TABLE 7-1 COUNTRIES HURT BY CONTINUED COOLING

Countries That Could Be Obliterated by Growing Ice Sheets[1]

Canada * ‡	Finland *	
Greenland	Switzerland † ‡	
Iceland	Liechtenstein	
Irish Republic	Nepal	
United Kingdom ±	Sikkim	
Norway ‡	Bhutan	
Sweden * ‡	New Zealand	
Denmark #		

Countries Threatened by Widespread Glaciation[1]

United States ±	Germany, FDR † ‡ and DDR * ‡	Bangladesh
Soviet Union ±	Poland #	China ±
(some parts obliterated)		Indo-China **
Mexico *	Austria #	Australia ‡
Colombia #	Afghanistan	
Argentina † ‡	China ±	
Chile ‡	Australia ‡	
Netherlands † ‡		

Zambia		
Rhodesia		
South Africa *† ‡		
Botswana		
Afghanistan		
Pakistan *† ‡		
India ±		

Countries Threatened with Severe Drought[2]

Mexico *	Ghana	Central African Republic
Guatemala	Togo	Congo
Guyana	Dahomey	Zaire #
Surinam	Niger	Kenya
French Guiana	Nigeria	Tanzania
Brazil *† ‡	Chad	Angola
Argentina *† ‡	Cameroun	Namibia
Uruguay #		
Paraguay		
Mauritania		
Senegal		
Guinea		
Mali		
Upper Volta		

[1] *Source:* Nigel Calder, *The Weather Machine* (New York: Viking Press, 1975).
[2] *Source:* Ibid. Calder is quoting a list by Rhodes Fairbridge.

Symbols: ± indicates nations already possessing nuclear weapons; * indicates countries projected to have significant installed reactor capacity by 1980, according to Dr. Mason Willrich, *Bulletin of Atomic Scientists*, May 1975; † indicates nations possessing the technology to build atomic weapons, but that have not ratified the nuclear Non-Proliferation Treaty, from Dr. Thomas A. Halsted, *Bulletin of Atomic Scientists*, May 1975; ‡ indicates nations listed as "candidates for the nuclear club" by William Epstein, *Scientific American*, April 1975; # indicates nations not otherwise marked that have reactors completed or under construction, according to Epstein, ibid.; ** indicates that Calder and Fairbridge list Indo-China as a drought-threatened region, and Willrich lists Thailand, which by some definitions is part of Indo-China, as having significant installed reactor capacity by 1980.

of its own), and the Dutch (rulers of the oil fields of Indonesia). Japanese warlords estimated that if they did not attack in December 1941, within three to six months their industries would grind to a halt and their people would have to beg for food. Japanese history to this day calls the embargo the "ABCD Encirclement."

Thanks to U.S. policies seeking an Asian superpower to counterbalance Chinese influence, Japan is at least "eight months pregnant" with nuclear weapons potential. The United States has supplied advanced missile and reactor technologies to Japan, and given her enough plutonium to make more than seventy atomic bombs. Japan has not yet ratified the nuclear Non-Proliferation Treaty. If she wanted nuclear weapons tomorrow, she need only, in the current phrase, "tighten a few screws."

Of the other nations threatened with indirect effects of climatic change—cutoffs of fuel and food—several could overnight become atomic weapon powers: Taiwan, Indonesia, South Korea, Iran, Spain, Egypt, and Israel. Of these, Israel is best equipped to fight an old-style nuclear war. Her U.S.-built bombers can ferry nuclear weapons 2,300 miles. Moscow is 1,700 miles from Tel Aviv, and Soviet ports on the Black Sea are less than half that far. South Korea has announced its intention to acquire atomic arms to defend against a North Korea that soon could suffer famine. Egypt reportedly has sought help in developing atomic weapons from China and India, and is now planning to build ten nuclear reactors.

Poorer nations will also be squeezed between the food and oil weapons of FOPEC and OPEC, and many will suffer famine and loss of national sovereignty. Under such circumstances, nuclear blackmail may seem their best option. Economist Robert Heilbroner anticipates many small nations getting atomic arms within the next few decades and using the weapons for blackmail and "wars of redistribution" of scarce resources. Such tactics, he wrote in 1974, may "be the only way by which the poor nations can hope to remedy their condition."

Even without atomic weapons, small nations can develop mega-weapons for terrorist uses. Germ weapons and nerve gas are examples. Civilized nations recoil at the use of such things—as well they might: urban units like New York City or Los Angeles, which supply millions of people with food, water, and fuel by a few fragile, centralized supply lines, make perfect terrorist targets.

But the governments that will arise in poor nations, faced with famine at home and exploitation from abroad, could be revolutionary, even fanatical. Would they shrink from using any weapon that won food?

Consider an extreme scenario: In 1983 drought is causing severe famine in a small sub-Saharan African nation. A coup d'état brings a fanatical colonel, Ami Kadaffy, to power. Kadaffy's philosophy is like that of the American terrorist "Weathermen," a cosmic interpretation of Karl Marx that sees Earth's white races as the "bourgeoisie," destined to be overthrown by the "proletariate" of colored peoples.

Col. Kadaffy wins some worldwide support by calling for a "climatic alliance" that will pit food-producing white nations of Earth's higher latitudes against colored nations nearer the equator. His views have a certain logic, for skin pigmentation is a climate-produced adaptation to sunlight. Too much sunlight can produce Vitamin D toxicity, but too little can lead to a deficiency. Thus pigment evolved as a protection for humans in the tropics, where the most sunlight shines and the atmospheric ozone layer is thin, and a lack of pigment evolved as a source of needed Vitamin D where less sun shines.

Kadaffy has weapons. In several equatorial countries he urges creation of factories to pump fluorocarbon gases into the air. He also supplies canisters of bromine gas to black dissidents in Europe, South Africa, and the United States. These canisters are attached to weather balloons, purchased from mail-order houses, and floated into the atmosphere. Twenty miles above the ground, the canisters release their gas.

Within a few months Soviet and American scientists report that Earth's atmospheric ozone layer is thinning, and that the resulting dose of solar ultraviolet radiation (see Chapter Two) is already causing increased skin cancer and other maladies among white-skinned peoples. Col. Kadaffy announces what he has been doing, and demands food and other political concessions from Earth's Caucasian nations as the price of ending balloon bombardment of the skies and shutting down the fluorocarbon plants. The U.S. and U.S.S.R. consider destroying the plants, but they cannot seize all terrorists with bromine gas balloons. They yield to Kadaffy's demands.

This frightening scenario is quite possible. Harvard University Atmospheric Sciences Professor Michael B. McElroy in June 1975 wrote that bromine gas, injected into the upper atmosphere by airplane, small rocket, or balloon, could punch a hole in the ozone layer. Such a hole would remain aloft above a given latitude while Earth turned under it. The hole could remain open for days or weeks, exposing all beneath it to dangerous doses of ultraviolet radiation. This, said Dr. Paul Crutzen of the National Oceanic and Atmospheric Administration, ''would become the poor man's nuclear weapon.'' ''It would be a doomsday weapon,'' said Dr. McElroy, ''because it would cause equal harm to friend and foe alike.''

In this last judgment McElroy is wrong, for the increases in ultraviolet radiation a thinning of the ozone layer would cause can directly do far more damage to Caucasians than to people with protective skin pigmentation. Thus, an attack on the ozone layer can be a race-specific weapon of war, and might seem an ideal tool to a racist fanatic—an apt answer to triage policies he sees as genocidal.

Even nuclear ''wars of redistribution'' could have similar effects. Air-burst atomic weapons would deplete Earth's ozone layer, and this, in addition to dosing the planet with various lethal radiations, could cause further upsets in our climatic balance. The use of any such mega-weapon threatens the whole web of life on Earth.

But before dismissing all future Kadaffys of the world as madmen, we should recall the reaction of leaders of the industrial nations to the Arab oil embargo and later OPEC policies. ''There comes a point where the conditions under which oil is supplied lose their commercial character and become issues of national survival,'' said U.S. Federal Energy Administration chief John Sawhill, speaking for the White House in late 1974. ''At that point —and we have long since passed it—we must explore the full range of options at our disposal to protect the national interest.'' Since he spoke, both the Department of Defense and Library of Congress have analyzed the worth of going to war if oil is again embargoed.

Facing famine as Earth's climate keeps fluctuating, many leaders of less developed nations will feel a similar rage as the climatocratic rulers of the planet threaten to embargo their food. Poor nations cannot be viewed as the helpless wounded of hospital triage wards. They may find the power themselves to determine who in the cooling world lives and dies.

8

Coping
with Climate

I
The Soviet Union

By late November 1941, Hitler's army was within sight of Moscow and at the gates of Leningrad. Snow covered the ground, but the winter was mild, as winters had tended to be since the 1930s. Soviet troops fought bravely, but the Nazis had control of the air and superior weapons. German victory seemed assured.

But in early December, just as Hitler's Japanese allies attacked the U.S. naval base at Pearl Harbor, the Russians gained an ally of their own, a merciless force they called "General Winter." Within days temperatures plummeted from zero degrees Celsius to minus 36 degrees. Nazi power froze in its mechanical tracks. Airplanes became unflyable. Armored tank turrets became unturnable. And German soldiers found their booted feet frozen, their fingers frostbitten. They fell back to winter positions. This was the first time Hitler's armies would be stopped.

"General Winter" had helped rescue Russia before. Another maniac bent on world conquest, Napoleon, led an army of 600,000 men against Moscow in 1812. Russian troops skirmished with him but refused battle. Instead they retreated, luring him deep into the country. The Russians let him capture Moscow, but it burned the night his troops arrived. Sheltered by tents, Napoleon and his army lingered in the city, waiting for the Czar's surrender. It never came, but winter did. On October 22 the French forces retreated for home, battling snow, flooded streams, and mud. Wagons broke down, horses died, men froze or died of disease and starvation. The retreating soldiers got lost in blizzards, and cut down by

129

the sabres of Cossacks who appeared, then vanished, in the swirling snowstorms. In mid-December what remained of Napoleon's *Grande Armée* staggered across the border into Germany. Scarcely 300,000 survived.

To Russians, winter is occasionally an ally but usually an enemy. The problem is simple: the country is too close to the North Pole, and too big. The border between the contiguous United States and Canada is at 49 degrees North Latitude. More than 60 percent of the Soviet Union's grainfields lie above this line. Moscow is as far north as Ketchikan, Alaska, and Leningrad is nearly as far north as Anchorage.

Russia also suffers what meteorologists call "continentality." Her farmlands are far inland from any ocean. Oceans, by keeping cooler than land in summer and warmer in winter, help moderate temperatures. But without this maritime influence to moderate their climate, Soviet grainlands are victims of extremes—terrible cold, devastating heat.

Her topography, too, is bad for food production. The Russian steppes are flat. Winds and storms rake the region, and no mountain or forest stops them. In winter the wind blows constantly, pushing away snow needed to water crops. In summer, high-pressure areas settle over thousands of miles of Soviet farmland at once, keeping rainclouds away and driving temperatures up. One such front stayed over the steppes for several weeks in 1972; weather turned so hot, reports Library of Congress analyst John Hardt, that peat bogs in the region caught fire spontaneously.

These forces of climate mean that the Soviet Union cannot reliably produce as much food as it wants, as the United States can. As former Soviet agriculture minister V. V. Matskevich has written: "In the U.S.S.R. only 1.1 percent of the arable land lies in areas with an annual precipitation of [28 inches], while in the United States it is 60 percent. . . . More than two-thirds of the area sown to grain crops in the U.S.S.R. is located in areas with insufficient precipitation. . . . Severe and very severe droughts occur once in three years. . . . Only about one year out of every three or four can be considered more or less favorable."

In the north, growing seasons are short; frost comes as early as September. The summer sun may stay in the sky until midnight in some northern regions, but sunlight is weak. If rain were dependable, some Soviet areas might produce wheat harvests to rival Canada, which shares the same latitude, but Russia lacks Canada's mountains and maritime climate. In many of her southern lands the Soviet Union is like a desert. Too cold in the north, too dry in the south, she has problems. In 1975, for example, odd air-pressure patterns produced a warm winter, with inadequate snowfall, in much of her prime grainland. When the spring plantings of corn and wheat began to sprout, a June heat wave parched the young plants. The result: a harvest short of its 215-million-ton goal by perhaps 80 million tons, the worst harvest in a decade, which necessitated purchase of grain from the U.S. and Canada.

Before this century, the peoples of the Russias had little choice: they endured the harsh climate. They prayed for good weather, and they used primitive climate modification. When spring neared, they would sprinkle dirt and ashes on snowdrifts that covered their garden plots. Clean snow might reflect away 90 percent of the sunlight striking it, but with a surface darkened by dirt it absorbs more than half the solar energy received. Russian peasants knew only that dirtying the snow made it melt faster, which meant they had more days to grow food on the land.

Following the 1917 Russian Revolution, a new religion, Marxism, spread through the nation. Among its dogmas: Man has replaced God. Nature is subject to human manipulation. With science and socialism anything can be done, for they are the keys to destiny. The new beliefs spawned rhetoric and beliefs reminiscent of the fictional character Superman, who could "change the course of mighty rivers, bend steel in his bare hands . . ." "We will change the surface of our planet," wrote Russian civil engineer Nicolai Romanov. "We will tame the oceans and force the great currents into other tracks. We shall even create artificial seas."

But for Lenin and all subsequent Soviet leaders, two overwhelming goals carried over from the days of the Czar: the Nation needed self-sufficiency in food and industrial production, and it needed seaports. Both had always eluded Russia, but without them the Soviet Union could never become a great, independent power. Perhaps by using science to change the weather and the climate these things could be achieved.

Today the Soviets spend more on weather and climate modification than any other nation. Most of their activities are secret, but they admit to a $30 million budget for research and modification of weather. They now routinely seed clouds with silver iodide and other chemicals. Inside the clouds water condenses around the chemical crystals, and sometimes this causes rain. The Soviets say that they have increased rainfall by 10 percent or more. If this is true, they do a lot of seeding.

Sometimes natural rain freezes, is caught repeatedly in updrafts, gains new layers of frozen water, and when heavy enough falls as hail. Hailstones annually do at least $200 million damage to crops and buildings in the United States. In August 1975, hailstorms in China were severe enough to "kill small animals in the fields" and to cut down wheat crops like a giant scythe. The Soviets claim they now can reduce hail damage by 70 to 90 percent.

They use rockets to suppress hail. The small rockets impregnate clouds with silver iodide. The result: more but smaller hailstones that are "mushy" rather than hard, and hence cause less hail damage to crops. The Soviets seed some hail-causing clouds from the air, spotting them by radar. More commonly, the state collective farms are equipped with rockets, which cost about $10 apiece and resemble American Fourth of July skyrockets; when storms threaten, the farmers shoot them off to defend their fields. In Yugoslavia the state farms are armed with cannon, and farmers fire silver iodide shells to suppress hail. They say it works, but National Science Foundation research into hail since 1972 has been unable to duplicate Soviet or Yugoslav successes.

The American Great Plains are flat and windswept, much like the Russian steppes. During the Dust Bowl drought of the 1930s, the U.S. Department of Agriculture planted tens of thousands of trees as windbreaks on the plains. In August 1975, Agriculture Secretary Earl Butz warned Americans to stop cutting these trees down; by blocking winds, weakening storms, and slowing evaporation the trees are each worth $128 yearly to American agriculture, Butz said. During the 1930s Soviet Premier Joseph Stalin ordered tree planting in the steppes. Today rows of trees continue to be planted there, along with rows of sunflowers. Mounds of dirt and snow fences are also being built to retard the endless wind. Windbreaks are one means of climate modification.

Lenin and Stalin realized early that the Soviet Union could grow all the food it needed if its southern lands had enough water. But where could water be found? The rivers that flowed into these areas were fed by melting glaciers. The Soviets set out to speed glacier melting, using methods Russian peasants had employed for centuries. In early tests coal dust and soil were spread by hand over glaciers in the Tien Shan. Later, airplanes dusted nineteen glaciers in the region, near the Chinese border, with five to ten tons of coal dust per square kilometer. The thin layer of dust increased sunlight absorption, and glacier meltwater rose by 55 percent. Channels were also blasted in several glaciers; this too speeded melting and increased runoff.

Of even greater interest to Soviet engineers was the water wasted in the north. Ten major rivers flowed north and dumped billions of gallons of fresh water each year into the Arctic Ocean. Under Stalin, plans were developed to divert this water south by changing the course of these rivers. By 1970 the Soviets had completed the world's largest hydroelectric dam near Siberia's Lake Baykal, blocking Angara River waters that once flowed into the Arctic. They also finished several reservoirs along the Volga River system, including the 100-mile long "artificial sea" of Tsymlyanskaya. Such bodies of inland water improve climate by keeping warm in winter, cool in summer, and by providing evaporation to enhance rainfall in fields nearby.

In 1971 the Soviets set off three atomic explosions, of 15 kilotons each, to help dig a seventy-mile-long canal from the Pechora River, which empties into the Arctic Ocean, to the Kama River, which flows south, joins the Volga River, and empties into the Caspian Sea. The level of the Caspian has dropped more than eight feet since 1930; the fish catch is falling, and the Soviets say new water is needed to "save" the sea. The water is wasted in the Arctic, they say. Soviet engineers are now planning to complete digging of the canal by using at least 250 more atomic blasts, mostly in the 100- to 200-kiloton range, to be fired 20 at a time. They believe, as did exponents of America's Project Plowshare in the 1950s, that this can be done without adding radioactivity to the atmosphere, soil, or water.

Trouble is, many of the experts who studied reversing Arctic rivers in Stalin's time talked of water for the south as a by-product of the scheme's bigger gain—an Arctic Ocean free of ice. The idea was promoted first as a means of climate modification, and only secondarily for the water it could bring southward.

Rivers like the Pechora carry fresh water into the Arctic Ocean. Fresh water freezes at a higher temperature than does water with salt in it. Thus, to simplify, the less fresh water that enters the Arctic, the less ice that ocean will have. Canada, more than a decade ago, considered reversing her Arctic-flowing rivers and thought better of it; such rivers carry warmth, writes Kenneth Hickman of the Rochester Institute of Technology, and diverting them would make the Arctic Ocean colder. Either way, a changed Arctic would drastically alter world climate.

The Arctic Ocean has been a great frustration to Soviet leaders, including the Czars. Russia has one of the longest seacoasts in the world, but it borders a sea that is frozen much of the year. The Czar warred with Japan in 1904, trying to win a warm-water seaport, but lost in 1905. The Soviets turned to battling the elements. As early as 1921, Lenin sent scientific teams into the Arctic to find ways to reduce its ice. Stalin established the first floating ice station for research during the 1930s. The Soviets used dust, rock salt, and waste oil on several occasions to melt ice floes blocking the White Sea approaches to their port of Archangel near Finland. Soviet climatologist Mikhail Budyko has proposed using aircraft to dust large regions of Arctic Ocean ice. Once melted, he writes, the Arctic would not refreeze in winter; the sunlight absorbed in summer by a dark ocean surface would warm the region enough to prevent new ice from forming. (In 1975 U.S. Navy scientists warned that a major oil spill in the Arctic could indeed cause widespread melting—and perhaps, in consequence, a new Ice Age, in accord with theories to be discussed in Chapter 13. Days after their warning, oil was accidentally spilled across several acres of tundra in Alaska.)

Budyko's plan might work, said Harry Wexler, the late research director for the U.S. Meteorological Service, but the odds are against it. The Soviets had dusted glaciers with up to ten tons of coal dust per square kilometer. To do likewise over the whole Arctic ice pack would require 150 *million* flights by cargo airplanes, each carrying 10,000 pounds of carbon black. "This," wrote Wexler, "would take considerable time," during which wind and storm would wash the thin film away from areas already coated. The task would never end. Budyko must have smiled to read this. Ice in the Arctic has been slowly shrinking since the 1920s, and Budyko merely proposed accelerating the process. The whole Arctic need not be coated; every inch of ocean that can be opened to sunlight increases the total heat energy in the Arctic Ocean and speeds the polar-ice retreat.

If the warming of Earth's climate between 1880 and the 1940s caused a retreat in polar ice, will the cooling that has since taken hold expand the ice once more? The Soviets think it possible, as evidenced by the new urgency in their plans to divert rivers, by renewed discussion among Soviet scientists of ideas to melt Arctic ice and control climate, and by the priority the Soviet Union now gives to the quick building of icebreakers.

During the 1950s and 1960s the Soviets were optimistic that between the Earth's warming trend (the cooling was as yet scarcely recognized) and their science, Russia would gain year-round ports on the Arctic Ocean. In 1967 the Arctic seemed clear enough of ice that the U.S.S.R. could announce its new pride to the world: the legendary Northwest Passage, the shortest route between Europe and the Orient that explorers had searched for since 1492, had been found, and the Soviet Union controlled it.

It seemed a bargain. By sailing through the Soviet Arctic, a ship going from London to Yokohama could save thirteen days' travel time and $36,000 in costs over what the trip required via the Suez Canal. Soviet aircraft would escort the ships, and Soviet icebreakers would protect them. The Soviets publicized their new route and made trial shipping runs across it. As analyst David Fairhall commented in 1971, "The reaction of British and Norwegian shipowners was distinctly cool. The Danes, Italians, and particulary the Japanese, were reported to be interested. But in fact nothing seems to have come of the Russians' ambitious plans . . . not a single foreign vessel [has sailed the route]."

Perhaps one problem, notes Fairhall, is that ice is an ever-present danger in the Arctic. Shipowners weighed potential savings against the real risk of having a valuable ship locked in Arctic ice for a winter and decided the gamble was not worth the gain. After the severe winter of 1971–72 the Arctic route seemed out of the question for all except Soviet freighters, which have reinforced hulls like icebreakers. As recently as late August 1975, a fleet of

American ships bound for oil fields in north Alaska were trapped in Arctic ice. The Arctic environment is hostile and uncertain year-round.

But what if the Arctic were ice-free? Some Soviet scientists dream of this possibility and see in it the answer to most of their country's problems. They believe it could make winters shorter and growing seasons longer, and that evaporation from the ice-free Arctic would increase rainfall throughout the country and give Russia a maritime climate that would moderate temperatures. Soviet ports on the Arctic Ocean would become centers of world shipping and commerce.

The price for this prosperity would be a warming climate for the planet, for Earth would absorb 2 percent more total energy each year from the sun if the Arctic were free of ice. With Arctic ice gone, the ice in Greenland and Antarctica would soon start melting. World sea levels would rise by about 400 feet when all the ice disappears, but the Soviets might find this acceptable. Studies by U.S. Department of Defense researchers indicate that of all the world's developed nations, the Soviet Union would be hurt least by such a rise in the oceans.

Soviet research continues, spurred by fear that if the global cooling continues the U.S.S.R. will suffer more than almost any nation, and by hope that if the Arctic can be freed of ice Russia will become at least as prosperous as the United States. Both the U.S. and U.S.S.R. hold vast territories bordering on the Arctic Ocean. Both could benefit if it became navigable, safe, and less stormy. Thus in 1972 the two nations began jointly studying weather in the Bering Sea, between Alaska and Siberia. By 1973 the United States, Russia, and Canada were cooperating in POLEX, the Polar Experiment of the Global Atmospheric Research Program (GARP). POLEX studies changes in Arctic ice with the seasons and how they affect weather. Such studies will find out how climate in the Arctic can and should be modified to best serve the three great powers sharing the top of the world.

9

Coping
with Climate

_____ II

The United States

Benjamin Franklin was the first American, and probably the first
scientist anywhere, to propose climate modification as a weapon
of war. During the American Revolution against British rule two
centuries ago, Franklin suggested that if Yankee ingenuity could
divert the Gulf Stream current flowing northward along the colonial coast, Great Britain would be brought to its knees by the
consequent cooling. The founding father of American invention,
Franklin had studied the Gulf Stream, recommended that American ships bound for Europe use its eastward motion across the
North Atlantic to speed their travel, and on his own voyages
analyzed its temperature. He was the first to recognize its role in
warming Europe, and the first to see the military potential in
environmental manipulation on a mass scale. His idea would have
worked, for a southward shift in the Gulf Stream accompanies
every Great Ice Age in Europe, but making it work was far beyond
the technology of 200 years ago.

In 1912 another American inventor laid out his plan for modifying the Gulf Stream to change world climate. The results he
predicted may have been wrong, but the methods he proposed for
changing the current were well within the realm of possibility then,
as now.

He was Carroll Livingston Riker, a lifelong native of New York

137

City. As an engineer he had invented floating torpedos used during the Spanish-American War, designed the giant dredge that filled in the Potomac Flats swamp in Washington, D.C., and served as an expert consultant during the building of the Panama Canal. Thus, although his idea seemed far-fetched, Riker was acknowledged to be a genius of sorts. When his book *Power and Control of the Gulf Stream* appeared in 1912, proposing that for a mere $190 million—less than the cost of the Panama Canal—the United States could gain many climatic benefits, he was taken quite seriously.

Riker's plan was logical and lucid. The Gulf Stream travels up along the American coast without any problem, he said, but when it turns east to cross the Atlantic Ocean it collides with the icy Labrador Current coming down from the Arctic. This collision in relatively shallow water weakens the Gulf Stream and cools it. But this would change, he wrote, if a simple jetty 200 miles long could be built from Cape Race on Newfoundland to a point just beyond the underwater Grand Banks. This jetty would keep the two currents apart until they reached the deep water of the North Atlantic. Once in deeper water, they would not collide, for cold water sinks to the ocean bottom and warm water stays near the surface and the Gulf Stream would pass unweakened over the sinking Labrador Current.

Off the tip of Greenland, according to Riker, the more-powerful Gulf Stream would divide. Half would throw increased warmth against Northern Europe, and half would thrust into the Arctic, flowing at least 400 miles farther north than ever before.

The benefits of this would be enormous, said Riker. Fog would disappear in much of the North Atlantic. Icebergs—one of which destroyed the "unsinkable" Titanic that year—would melt before they reached these new rivers of warmth within the ocean. And most important, within a few months all the ice in the Arctic would melt.

The melting of the Arctic would improve world climate in two ways. Without ice to their north, Europe and North America would be freed of chilling storms and icy ocean currents; they would warm. And without North Polar ice, the surviving ice pack at the South Pole would become the heaviest part of our planet. Centrifugal force would then tip the Earth, for the South Pole would tend to stay farther from the sun than the rest of our planet.

With the Northern Hemisphere tipped more toward the sun, Europe and North America could expect warmer climate. Riker said nothing about a rise in world sea levels if the Arctic and Greenland ice packs melted; perhaps he assumed the water would be taken up in Antarctica, which with less sunlight would certainly get colder.

Today experts think this last conclusion absurd; the world would not tip significantly from a melting of Arctic ice. His other predictions seem much more reasonable, although the extent of their consequences is doubtless much greater than Riker believed. For example, what would happen if the Labrador Current flowed unimpeded, pouring its cold water across the Mid-Atlantic and into the tropics? Even now scientists are unable to answer such questions clearly.

But in 1912 Riker's plan got tremendous support. Such notable companies as R. J. Packard, Harper & Bros., Harvey Fisk and Co., and others petitioned Congress to provide money for a study of the Labrador Current. *Scientific American* offered to furnish a team of experts for such an expedition. The head of the New York branch of the Hydrographic Office said Riker's idea was feasible. The Port of New York asked Congress for money to study building Riker's jetty, and the New York Board of Trade and Transportation lent its support to such investigations.

Some criticism was heard. A *New York Times* editorial fretted:

> Mr. Riker does not say what will happen in northwestern Europe after his big dike has been built. Perhaps he does not care. . . . The Britishers, after all, are human, or almost, and we have always been told that if the Gulf Stream didn't run just where it does now, their present weather, which is none too good, would become intolerable.

Riker replied that weather in all of Europe would improve because of his jetty.

Surprisingly little criticism was raised over how the jetty was to be built, perhaps because its ingenuity charmed and distracted would-be antagonists. At first Riker proposed building a very thin wall along the 200 miles. Later, this was modified to the building of a few key pillars with underwater cables stretched between them. In both cases the genius of the idea was that the Labrador

Current would build the jetty itself, by depositing sand against some sort of barrier. In his later proposal, a special cable would be designed to catch sand while resting just atop the sand already accumulated. Within two years the accumulation of sand behind the cable would be blocking more than two-thirds of the Labrador Current, he said. Would this have worked? Riker's idea came close to being tried, but fate worked against it.

A draft bill authorizing $100,000 for study of Riker's jetty was put before Congress in 1913, but it was killed in committee. War was brewing in Europe, and tampering with the Gulf Stream seemed too much like political involvement to isolationist American lawmakers. By then the idea was commonly discussed by Americans in saloons and salons, often as a way to wreck climate in those so-and-so European kingdoms. As a technological idea the jetty seemed too uncertain in its results. As a political idea it seemed too hot to handle. When World War I came it was forgotten.

But Congress was willing to fund other environmental tampering. As early as 1891 it appropriated $9,000 for rainmaking experiments under control of the Department of Agriculture. That year tests were done in Washington, D.C., and later in San Diego, Texas. Dynamite was hung from hydrogen-oxygen balloons, and the U.S. Weather Bureau at San Antonio reported that the explosions apparently caused some rain.

Few nations have better climate than the United States. Turn the country upside down and it resembles Egypt, where a fertile river running through the heart of the land created a cradle of civilization. In the heartland of America the ancient Mississippi River has created topsoil 20 feet deep in some places, the richest land on Earth. But the U.S. is further blessed. Egypt's river flows into the Mediterranean Sea to the north; the Mississippi flows into the Gulf of Mexico to the south. Warm moisture-bearing air moves from the equator toward the poles of our planet—thus Egypt is a desert swept by hot dry air from the south, but the U.S. Midwest gets abundant rainfall in most years from water evaporated in the Gulf of Mexico and blown north. America's East Coast gets ample rain in clouds blowing inland from the Gulf Stream Atlantic regions. The Pacific Northwest gets up to 400 inches of rain annually in a few spots where moisture-bearing air from the tropical Pacific

Ocean is channeled by the Rocky and Sierra Nevada/Cascade Mountain ranges into oncoming cold fronts headed south from Alaska. Such collisions make the warm air give up its moisture.

If you could build a country from scratch, moving mountains and rivers and seas as you wished, you could scarcely create a more ideal climate over so large an area—with a few exceptions. In the Southwest, in Nevada and Arizona and New Mexico and Utah, the air that travels northward from Mexico is generally dry, and these states feature millions of acres of virtual wasteland. Occasional droughts and extremes of temperature hit the Midwest, and in some regions like North and South Dakota can come often. During the peak of the recent warming trend, in the 1930s, severe drought ruined agriculture in Oklahoma and Arkansas. But for the most part the United States enjoys classic maritime climate. Temperatures are moderated, and rainfall is kept plentiful, by oceans to the east and west, by a tropical sea to the south, and by a chain of Great Lakes in the north.

But when the rains fail, or when storms come instead, Americans have tried to change their weather, often in strange ways. Indians in the Southwest resorted to the magic of rain dances, and may have used a method missionaries observed elsewhere in the Americas in the 1700s—special fires that added huge smoke clouds to potential rain clouds, a technique somewhat like those used in modern rainmaking. Anglo pioneers used prayer, but many believed gunfire could prompt rain. Skyrockets and other explosives used during annual Fourth of July celebrations, folk wisdom held, often caused rain, but this appears to be more fable than fact.

The history of the United States, like other nations troubled by droughts, includes myriad professional rainmakers who for a fee would use assorted hocus-pocus to conjure precipitation (or, also for a fee, to prevent it). Among these are the legendary Charles Warren Hatfield, who in 1916 was hired by the city of San Diego, California, to end a drought. Hatfield set up his 25-foot-high platform and started his strange machine that belched out smoky chemical fumes. The city had guaranteed him $1,000 for every inch of rain that fell. Downwind from his apparatus 20 inches fell in nearby country, washed away a dam, and caused a flood that killed seventeen people and did millions of dollars' worth of damage. San Diego offered to let Hatfield flee town unpaid or accept responsibility for the flood. He fled. Another rainmaker was Dr. Wilhelm Reich, the eccentric disciple of Sigmund Freud who claimed to heal people with "orgone" energy—"sex energy," some critics called it. Before he was sent to prison in 1957 for selling orgone devices, which the Food and Drug Administration deemed fraudulent, Reich demonstrated a device he called "cloudbuster." A few followers say they saw Reich aim this device, which he also named "an orgone cannon," into the sky, and watched as rainclouds formed where the device pointed. Reich died in prison. His work has largely been ignored until recently, when some Soviet scientists have taken a passionate interest in his devices.

The worth of some primitive rainmaking techniques is by now well-known. In 1891, for example, Chicagoan Louis Gathmann was issued a U.S. patent for a device that released liquid carbon dioxide in the sky via an exploding balloon. In 1946 a young researcher at General Electric Corporation in Schenectady, New York, demonstrated that dropping solid CO_2, dry ice, into clouds would produce snow and rain. Vincent Schaefer discovered in July of that year that dry ice could generate snowflakes in a misty freezer. In November he dropped some dry ice from an airplane over Greylock Mountain in Massachusetts, causing a brief snowfall. General Electric lawyers ordered Schaefer never to do such cloud seeding again, lest the company be sued for storm damages. But out of his experiment came proof that humans could create rainfall, and came military contracts for weather-modification research under Project Cirrus. The project quickly confirmed Schaefer's findings, and discovered too that dry-ice seeding could clear clouds and fog where used. Unlike General Electric, the government was legally immune from lawsuit (at that time) if it caused a harmful storm.

In 1948 another General Electric scientist, Dr. Bernard Vonnegut, discovered that snowflakes could be grown around the crystal structure of silver iodide. Using a smoke generator to burn the chemical, Vonnegut demonstrated that snow formed naturally around such crystals in cold air. Because air is cold at several thousand feet altitude, silver iodide particles can produce snow that melts into rain while falling. Dry ice then cost five cents per pound. Pure silver iodide cost $20 per pound. But the iodide needed no airplane or balloon to take it aloft. Project Cirrus tested silver iodide burners on clouds near Albuquerque, New Mexico, in July 1949. Within two days, more than an inch of rain fell over a wide area from Albuquerque to Santa Fe. The official estimate was that 480 billion gallons of water had dropped from the sky.

These and future experiments were carried out by one of cloud seedings great early advocates, General Electric scientist Dr. Irving Langmuir. He estimated that with efficient silver iodide burners the cost of cloud seeding could run as low as $1 for 1,000 square miles. Ideally, he said, seeding could double rainfall in the whole United States for only about $200 per week. Soon cloud seeding became a popular political topic of discussion. President Dwight Eisenhower appointed a committee to study the potential benefits and dangers of tampering with weather; in 1957 it reported that seeding could not double rainfall, but could increase it over the United States by between 9 and 17 percent.

But research funds for Project Cirrus were cut back, and the U.S. Weather Bureau showed little enthusiasm for cloud-seeding research, apparently because their experiments showed poorer results than those of the military. From 1957 through 1967 the government investment in peaceful weather-modification research was small.

In the private sector, however, rainmaking was becoming big business, despite the risk of lawsuits. During its past twenty years of operation, for example, Dr. Irving Krick's cloud-seeding company has routinely won more than $1 million annually in contracts for rainmaking. The firm, now located in Palm Springs, California, produced snow for the 1960 Olympics in Squaw Valley, made rain over farmlands in Algeria, Israel, Syria, the West Indies and other nations, and increased mountain rainfall to improve hydroelectric dam water supplies in Washington, Italy, France, and Spain. Denver, Dallas, Shreveport, Louisiana, and Greensboro, North Carolina, are among the cities hiring Krick and his silver iodide generators. He claims he can increase precipitation by 13 to 15 percent in most areas, and sells his services for less than $2.50 per hectare per growing season to farms and communities, a price he describes as "extremely inexpensive crop insurance." His is one of more than a dozen major companies selling weather modification in the United States.

Krick has been sued once, by a landowner who claimed his seeding of clouds near Oklahoma City in 1953 caused a flood. A jury found Krick not guilty of any negligence in his operations. Of the suits directed against weather modification companies, none has ever resulted in damages. One 1956 case in Washington state temporarily enjoined a company from using cloud seeding to suppress hail, another potential use of the process.

Between 1967 and 1972 cloud seeding was used by the Department of Defense against enemy supply routes in Southeast Asia. It proved a small success: rain muddied a few trails and flooded some food-growing areas. Rainmaking rose in the esteem of government agencies when its warfare use was made public. In 1972 the U.S. government began funding rainmaking on a large scale through the Department of the Interior's Bureau of Reclamation and its Project Skywater. As of 1975 the federal government was involved directly in nine weather-modifying programs. In the United States sixty-six nonfederal programs were underway. Twenty-two states and Puerto Rico have tried changing their weather. The leader among these is South Dakota, which now annually spends more than $1 million on cloud seeding. The price to farmers for these programs typically runs a few pennies per hectare. The U.S. government now officially spends roughly $26 million each year on weather modification, second in such spending only to the Soviet Union. At various times this modification has aimed at specific conditions: in 1966 the government began intense hail-suppression research with Project Hailswath in South Dakota, where seeding is still used against hail; in 1952 it undertook lightning suppression under Project Skyfire, in large part because lightning starts up to 15,000 forest fires each year, at least nine times the number set by humans. The efforts at hail and lightning suppression have had limited success.

Each year tornados rip across the U.S. Midwest. These "twisters" are fast-spinning funnels of air that leave narrow paths of devastation wherever they move. The yearly number of tornados has been increasing since the cooling began in the 1940s. In 1973 a record was set: more than a thousand struck in the Midwest and East; ninety devastated several Eastern states on April 4th alone, killing many people.

Recent research shows that tornados are caused by masses of cold air bumping into one another near thunderstorms. Once a swirling motion begins, their own suction keeps them going. But how to stop them nobody knows. An 1887 patent was granted in the U.S. for an anti-tornado explosive towns could mount on a pole to their southwest, from whence the storms would come. But tornados are tremendously powerful, like condensed hurricanes, and modern study indicates that only an H-bomb on the pole could reliably stop them—and it would probably stop the town too. The number and intensity of tornados keeps growing, as does their range. Twisters have been sighted in all of the continental United States. In 1975 they did vast damage deep in Canada as well.

By far the most destructive storms that touch the United States are hurricanes. Every year between June and November, the warming of the Northern Hemisphere creates a high-pressure area in the mid-Atlantic Ocean. This interferes with the usual flow of global winds and causes whirlpools in the sky, which then drift westward toward the Caribbean and the Gulf of Mexico. Sometimes a whirlpool gathers enough energy from sea water evaporating under it to become a cyclone, or even a hurricane up to 100 miles in diameter with swirling winds that can reach 200 miles per hour. A hurricane can continue for days, releasing energy equivalent to one hydrogen bomb per minute or more. An average year sees six hurricanes emerge this way and thunder west. Some strike islands in the Caribbean Sea, others swirl into Mexico or Central America, and some come to the United States. Every year, estimates the National Oceanic and Atmospheric Administration (NOAA), hurricanes "will kill 50 or 100 persons between Texas and Maine and cause property damage of more than $100 million. If it is a worse-than-average year, we will suffer several hundred deaths, and property damage will run into billions of dollars."

As early as 1947, American scientists were devising ways to control or weaken hurricanes. In that year General Electric scientist Dr. Irving Langmuir, who later would win the Nobel Prize, persuaded the military to use Project Cirrus airplanes to seed hurricanes with dry ice. This, he predicted, would force such storms to surrender the heat that drove them; they would weaken and harmlessly dissipate.

In mid-October a hurricane in the Gulf of Mexico turned and rolled across Florida. Back in the Atlantic, it headed north. When three Project Cirrus bombers reached it, the storm seemed dissipated already, but they dropped 180 pounds of dry ice into the walls of its spiral before heading home. When seeded, the storm was about 400 miles off the Georgia coast. Immediately after seeding, it began to gain strength and turned west. The reinvigorated hurricane hit Savannah and did $5 million in property damage. Langmuir accepted credit and blame, claiming the seeding had modified the storm. Critics said past hurricanes had done erratic things too, that the seeding might have done nothing.

In 1950 Dr. Langmuir proposed hurricane seeding with silver iodide, but by then government enthusiasm for such risky experiments was diminishing. Not until 1961 did military aircraft drop silver iodide generators into a hurricane. The storm weakened for several hours, and interest in seeding the giant storms revived among scientists.

New experiments began in 1962 under a new NOAA program, Project Stormfury. New rules were imposed as to which hurricanes could be seeded. When the project folded its tents temporarily a decade later, it was under orders only to experiment with storms "predicted not to come closer than 50 miles to a land area within 18 hours after seeding."

Such rules kept the infant program on the defensive. In 1965, for example, her bombers flew against Hurricane Betsy. But Betsy dipped and swerved, turning toward New Orleans, then cutting back across the tip of Florida, then recircling and at last running aground over Louisiana. "Bear in mind that this hurricane was not seeded," wrote University of Arizona meteorologist Louis J. Battan. "If it had been seeded . . . many people would have insisted that the sharp changes of direction were caused by the seeding. Hurricane Betsy is estimated to have caused 1.4 billion dollars' worth of damage. Had it been seeded . . . the courts would, in all likelihood, have been beseiged by lawsuits alleging that the seeding caused the effects and hence that the seeders were responsible for the damage. The thought of such a possibility must send shivers through the scientists of this project, but the research goes on.''

Of course some critics charge that Project Stormfury aircraft have seeded every destructive hurricane, including Betsy. Such charges explain why the project has seeded fewer than half a dozen storms in twice as many years. The federal government could claim immunity from such lawsuits under privilege retained from the days of kings, but the political cost of doing so would be high. This is one reason Congress has never passed legislation specifically authorizing Stormfury to modify hurricanes. No politician wants to accept direct responsibility for the harm a seeded storm might do. Thus, if the research goes on, it goes at a snail's pace. ''If hurricane modification is ever to be achieved,'' writes Battan, ''it must be possible to conduct research without risk of financial disasters to the investigators.''

Other means of weakening hurricanes have been tried. The storms are fueled by the energy released from water evaporation on ocean surfaces. But in the path of a hurricane seas can be coated with anti-evaporation chemicals like hexadecanol, a fatty substance derived from the heads of sperm whales. The United States has used hexadecanol coatings to suppress fog in the Panama Canal Zone, and the Soviets have used similar chemicals to suppress fog in the Arctic, both with great success. Could such sea coatings weaken or deflect hurricanes? Some experts say no, that the winds that sweep 200 miles from a hurricane's calm center would blow the film away before the main storm arrived. But reportedly it has been tried, with some small results. As early as 1958 the U.S. Weather Bureau also planned experiments to divert hurricanes with the heat updrafts from ''patches of burning fuel oil poured on the sea at crucial spots.''

The global cooling has somewhat reduced the number of hurricanes striking the United States, but this may be its only blessing. (In Florida this is a mixed blessing because hurricanes bring needed rain.) Tornados are on the increase. Rainfall is less reliable. Drought has intensified in the Southwest. Flooding has increased in the Mississippi Valley and the Northeast. Lawmakers who led the fight against U.S. rainmaking in Southeast Asia in

April 1975 urged the president to fund bigger peaceful weather-modification programs in the United States, both to improve water supplies for the nation and to reduce the need for energy by moderating climatic swings of temperature and storminess.* We can expect vast increases in U.S. weather- and climate-modification research in the near future.

*See Appendix Five.

10

The Lawless Sky

In April 1973 the Central American nation of Honduras officially accused the United States of theft. What had been stolen? Rain. According to the Honduras Meteorological Service, the U.S. had "upset the balance of nature" by using cloud seeding and other techniques to modify hurricanes near Miami, Florida. For centuries moist Atlantic air has been carried by winds across the Gulf of Mexico, producing rain when stopped by mountains in Honduras. But now, said the Hondurans, the U.S. Stormfury Project was squeezing the moisture from this air. Thus the United States was responsible for the severe drought they suffered.

The protest seemed like the rerun of a movie that appeared in 1949, at the dawn of the global cooling trend. Drought struck the Sula Valley of Honduras. Between January and August of that year only 5 inches of rain fell in a region accustomed to as much as 100 inches. Grass withered, and cattle died of starvation.

But in a third of the valley rain kept falling, the third controlled by the U.S.-based multinational United Fruit Company. The reason: the company had a small airplane and a professional cloud seeder dropping dry-ice pellets into clouds over their plantation. In an article titled "Rustlers in the Sky," *Time* Magazine gave details of the native reaction, which included accusations of rain theft and cartoons depicting the company's Texan pilot herding clouds with a lariat.

To make peace, the United Fruit Company offered either to stop its seeding activities or to seed clouds over the entire valley. Local cattlemen chose rain, and a week later the first seeded clouds dropped an inch and a half of water, enough to seem a downpour on the hard-baked land. Pilot Joe Silverthorne was delighted. "Say the word," he bragged, "and I'll flood the country."

Twenty-four years later, as the Hondurans assailed Stormfury, the project was pulling up stakes to move to Guam in the western Pacific Ocean, where it was to begin seeding typhoons in 1976. Already the Japanese government has protested possible rain theft by Stormfury operations. Japan relies on typhoons to bring rain, and has asked the United States only to experiment with storms that will not reach land.

Thus Honduras got its way, even if in 1973 it got no help with rainmaking. Ironically, in September 1974 the small country was battered by one of the worst hurricanes in its history, FiFi, which caused 10,000 deaths and billions of dollars worth of damage. FiFi was precisely the kind of storm Project Stormfury was trying to control.

But was FiFi's devastation an accident of nature? No, charged the director of the geographic research center of the University of Mexico in July 1975. Jorge Vivo said he had evidence that the United States used weather modification that "artificially detoured" the hurricane to Honduras to save Florida's tourist industry. Thus, said Vivo, the U.S. should accept moral and financial responsibility for the disaster. The agency in charge of Stormfury, the National Oceanic and Atmospheric Administration (NOAA), discounted these charges and said the United States had seeded no hurricanes since 1971. NOAA did not comment on whether or not other techniques the U.S. has tried in modifying hurricanes—coating sea surfaces in a storm's path with burning oil or with the anti-evaporation chemical hexadecanol—had been used against FiFi.

Hurricane FiFi destroyed most of the 1974 corn crop in Honduras. Drought returned to the small nation in 1975; by August it had killed 80 percent of the corn, and 700,000 Hondurans faced famine.

Ideally the Hondurans deserve a day in court, as do their neighbors in drought-stricken El Salvador who have made similar charges against the U.S., a chance to present any evidence they have that the United States has stolen their rain or directed a hurricane against them, a chance to gain justice for the wrongs they believe they have suffered. But this cannot happen, for at present

there is no international law of any kind governing weather modification. In 1974 the United Nations General Assembly resolved
that environmental warfare be forbidden, and a draft U.S.–Soviet
treaty banning such warfare was put before the thirty-one-nation
Disarmament Conference in Geneva in August 1975. Neither of
these actions have the force of international law, however, and
even if they did their application to Honduran complaints is far-
fetched. The U.S. and Honduras are not at war, and the
U.S.–Soviet proposal excludes peaceful uses of weather modification.

But if law existed in this realm, Honduras would have a difficult
time proving damages. The above cases illustrate some of the
bigger problems future international lawmakers will have in regulating global weather and climate controllers.

Question One: Who owns the skies? Centuries ago, most jurists
held that the private owner of a plot of land controlled its soil to the
core of the earth, its sky to the farthest edge of the heavens. He
could shoot a goose flying overhead, or tap an underground lake
beneath his claim.

With the advent of aircraft, new legal doctrines evolved to
discourage would-be gunners who might shoot down airliners
above their land, or charge tolls. Today's laws give to nations, not
individuals, control of the "airspace" above their territory—a
vague concept that includes most high-flying jets but excludes
satellites. It seems to mean control to the edge of the atmosphere.
And with large cities and ranches have come elaborate new doctrines of water rights, which apportion how much water from
adjoining rivers, lakes, and underground water tables a property
owner can claim.

With the emergence of new technologies of weather and climate modification, new legal doctrines are needed. What of those
"rivers in the sky," the moisture-bearing winds on which nations
like Honduras depend for their water? No international law protects nations that must share one river, but almost all have mutual
agreements to define their water rights. One such agreement
between the United States and Mexico, for example, provides for
both the amount and the saltiness of Colorado River water the
U.S. lets flow into Mexico. As yet no similar agreements exist
between any two nations sharing a wind route. Thus nations

assume the right to do as they wish with the skies over their own territory, including cloud seeding to squeeze moisture from passing clouds and, by extension, the creation of storms to prompt such precipitation—even though the storms may drift across international boundaries.

Question Two: What constitutes natural weather and climate? Honduras has suffered several severe droughts since the global cooling began three decades ago. Perhaps the cooling is the cause. Honduras lies on precisely the same wind and latitude belts as the Sahel region south of the Sahara Desert in Africa, which has been hit by droughts, because, say experts, the cooling has forced world monsoon rain zones toward the equator. As the map in Chapter Three shows, this Intertropical Convergence Zone used to lie against Honduras' west coast, but it shifted far to the south as the cooling deepened. Is this, and not U.S. weather modification, the reason for less rain in Honduras? This is possible, but with present scientific understanding of weather it is unprovable. It raises serious doubt, however, about U.S. guilt.

Weather and climate have changed drastically over millions of years without human intervention, and human ability to influence great natural forces is so small that we must assume that nature shapes almost all such changes today. Nature gives no nation a guarantee of permanent good weather or climate. But, to take a step back from Honduras' immediate cases, have certain industrial nations inadvertently caused the global cooling by their pollutions, or deliberately set it in motion with environmental tampering intended to improve their own weather or climate? No law as yet holds them responsible for such things, and science cannot yet establish clearly how human activities influence climate. But more than sixty nations have carried out weather-modification experiments thus far, and the United States is by no means the only country charged with robbing rain from its neighbors. Rhodesia and Israel have both been accused of rain rustling by bordering nations.

We are beginning to think of weather and climate as unnatural, as forces subject to human control—and hence as subject to human laws.

Question Three: What are a nation's weather rights? If Honduras got 100 inches of rain in 1945, was it entitled to 100 inches of rain in 1975? By old concepts of law the question is absurd. If the river by your farm dried up from drought, or if lightning hit your barn, these were "acts of God," and were excluded from consideration by both public servants and private insurance policies.

But today that is changing. The drought might be caused by another nation's cloud seeding, and the lightning storm might have started in a foreign weather-modification experiment. As international lawyer Edith Brown Weiss of Columbia University and the Brookings Institution told the American Meteorological Society in 1972, international law will soon be forced to evolve doctrines of "weather rights."

This will require a sophisticated rethinking of what "natural weather" means. Is good weather a limited resource in a worldwide balance of good and bad weather? If so, then if Rhodesia uses cloud seeding to get more rainfall for itself, somebody else must pay by suffering less rainfall. If so, justice may require some way of redistributing this resource, which for most nations makes the difference between feast and famine.

On the other hand, the potential for good weather may be infinite and merely in need of development and exploitation. As German climatologist Hermann Flöhn points out, the skies above the Sahara Desert contain as much water vapor as those over Europe's best farmland, but the water does not fall over the Sahara. If these African skies could be squeezed, perhaps the deserts would bloom from the rainfall, but without hurting rain elsewhere. If so, then perhaps all nations are entitled to conduct weather modification, and poor nations should receive weather-modifying free aid as they now do food aid. No nation should be forced to suffer drought if a bit of environmental tampering can prevent it. And morally no nation should suffer drought if the slightest possibility exists that weather modification elsewhere has deprived them of rain that would have fallen naturally but was removed from clouds headed their way.

The United Nations may eventually develop a system of world weather welfare, using technology to correct deficient weather and climate as needed and enhancing them where desired. The alternative is a world of increasing drought, as the cooling continues, in which injured nations will rightly or wrongly blame all their weather woes on countries with big weather-modification programs, like the U.S. and U.S.S.R., or on neighboring countries that try rainmaking methods. Even if untrue, politicians may find it expedient to accuse other nations of weather modification if the alternative is to accept blame themselves for failed policies. This would increase suspicion and hostility among nations and nurture the seeds of conflict that already exist.

Weather modification conducted even-handedly by an international organization like the United Nations would arouse less hostility than would national programs, which often come with political strings attached to even the most humane offers.

The same applies to negative weather: destructive storms. If a United Nations operation, rather than Project Stormfury, had been in charge of all hurricane-modification activities on the planet, the accusation that the United States steered FiFi into Honduras would have been groundless. If we suppose the worst—that the U.S. did somehow deflect this storm—we must still assume it was an act of self-defense rather than war. America's problem was that she could redirect the storm when over her own territory or international territory (the zone of operations for Stormfury was carefully mapped in the Atlantic Ocean), but she could not fly cloud-seeding flights over the airspace of Cuba or other Caribbean nations. A U.N. force could gain permission to fly missions anywhere and use weather modification to defend all nations equally from storms, as best it could. This would entail risks, and in the foreseeable future might at best mean directing a dangerous hurricane away from population centers but into small villages and islands. Responsibility for the destruction this would cause could be shared by the community of nations.

Likewise all attempts to modify Earth's climate should be done under international control, with international responsibility. Two climatologists at the National Center for Atmospheric Research, William W. Kellogg and Stephen H. Schneider, propose "No fault climate disaster insurance," the principle that when a majority of nations approve attempts to change the world's climate they accept responsibility to compensate any nation whose climate deteriorates following the attempted modification, whether or not blame for the change can be proven. It is sufficient that the nation to be helped has suffered unusual climate and weather conditions.

Such compensation could come in the form of money. More likely food aid would be provided as "famine insurance."

Given the present political problems of the United Nations, the day of global weather welfare or storm protection conducted by an international agency is many decades away. It will come when the nations now leading in weather-modification techniques realize that such programs are in their own best interest. The day may soon come when global weather forecasts are incessantly upset by national weather-modification practices, when droughts increase the likelihood of war throughout the world and the amount of hostility directed at the superpower nations, and when the threat of climatic disruption are all recognized as problems we share alike. In the meanwhile, some nations are finding good harvests at home and poor harvests abroad both in their national self-interest.

Question Four: Can international law be enforced? Not likely at present or in the near future, for the problems are many. The nations now engaged in weather modification are serving their own self-interest. They will not relinquish the right to carry out such "peaceful" activities. Thus, U.S. undertakings like Storm-fury will remain legal. Imagine a case brought before an international court, Honduras asking damages from the U.S. for Hurricane FiFi. Can Honduras prove U.S. planes seeded or otherwise tried to modify the storm? This could be difficult; judges may be forced to take the United States' word for what it did or did not do. If the storm was seeded, did this influence it or not? No scientist today could answer with absolute assurance; hurricanes do crazy things naturally, and weather-modification technology is yet primitive. What if more than one nation seeded the storm, say the U.S. and Cuba? Must weather-modifying nations accept full responsibility for a natural storm, or partial responsibility—and if the latter, how much? The storm might have hit Honduras without seeding. A case could even be made that the storm was Honduras' fault: her drought heated the land, which caused an updraft, which drew the hurricane toward her. The argument seems strange, but weather scientists might deem plausible. The issue before the court would be which expert to believe, for among themselves the experts disagree widely. Some say seeding has never influenced a hurricane in the slightest.

The above points, only a few among hundreds, suggest the complexity involved in enforcing weather-modification laws. Laws governing climate modification, an infant science, would face ten times the difficulties inherent in weather modification.

Because clouds, winds, and storms ignore national boundaries, we have considered the international legal problems of countries tampering with climate and weather. But law is evolving within countries too, and is establishing principles that someday may extend to international law. A case now in American courts may set a major precedent for all future regulation of the now-lawless skies.

On June 9, 1972, a small cloud-seeding airplane took off from the airport at Rapid City, South Dakota. A few miles west of the city its pilot dumped 650 pounds of salt, sometimes used to act as condensation nuclei for atmospheric water, into a bank of clouds. The plane was flown as part of the U.S. Department of the Interior's "Project Skywater," an operation to increase rainfall.

A rainstorm followed, and with it one of the worst floods in American history. There were 238 people killed, and hundreds of millions of dollars' worth of property was destroyed. Rapid City, a town of nearly 50,000 people, was almost washed away. State and federal governments provided $64 million in relief to help rebuild the city.

But this was not an "act of God," said some townspeople. It was caused by government cloud seeding. In 1974 flood victims asked the Interior Department to accept responsibility for their damages. The agency refused. On June 3, 1975, relatives of four people killed in the Rapid City flood brought suit against the Interior Department, seeking $1.7 million for loss of life and property damage. Twelve hundred of the flood's victims are considering a class-action suit, reportedly to ask $600 million in damages from the government. In both cases attorneys will contend that the flood was not wholly "natural."

In the whole history of weather modification only fourteen suits had been brought before this, says Howard Taubenfeld, a law professor at Southern Methodist University and America's foremost expert on weather law. Moreover, no damages have ever been assessed. Most dealt with weather modification under other laws: water rights, invasion of privacy, trespassing, destruction of property.

The Rapid City suit charges that the U.S. government was negligent and irresponsible in its cloud seeding, and it may have a case. David Hacker of the *National Observer* reports that evidence may show the cloud-seeding pilot on that June day was told to seed clouds *south* of Rapid City because officials feared a flood if clouds were seeded west of the town.

Even if the pilot seeded clouds contrary to the warning of his superiors, this does not prove his seeding caused any rain, or that the rain it might have produced was the flood's cause. However, a NOAA study said the Rapid City flood "averaged about *four times* the six-hour amounts that are expected *once every 100 years* in that area." And Project Skywater operates on the assumption that seeding works and can increase rainfall by 10 percent or more. Flood victims may decide that the cloud seeding was responsible for 10 percent of the damage they suffered. The suit will be a battle of experts, and is expected to be in the courts for years. Many specialists concur with the head of South Dakota's weather modification program, Martin Schock, who says "the cloud seeding did not materially contribute to the flood," but concedes that it might have increased rainfall somewhat. Did weather modification increase rainfall enough to warrant granting the victims damages? Most states, including South Dakota, license or otherwise restrict cloud seeders, and in this case the pilot was flying for a government project. If the court finds in favor of the flood victims, a new era of weather- and climate-modification law will have begun. In the meanwhile the Federal Government has started requiring environmental impact statements from all cloud-seeding operations.

Lawsuits and bureaucratic rules provide some restraint on weather and climate modification. Another form of regulation is direct democracy, as Colorado citizens showed in 1972. In Colorado's San Luis Valley, the people raise cattle and grow potatoes and, for the Coors Beer Company, Moravian barley. Coors invested heavily in cloud seeding over the valley—to provide the right amounts of rain while the barley grew, to prevent rain during harvest time, and to suppress hail that could harm crops. Farmers not growing barley in the valley objected that such weather modification robbed them of rain needed for their crops. Protests arose. A cloud-seeding airplane was shot at, and a dynamite bomb was used against a radar station that assisted in the seeding. The farmers conducted a straw ballot along with the regular November elections in 1972 and voted three to one to stop the cloud seeding. Coors threatened to cut off all barley buying from the valley if its cloud-seeding operations were stopped, but following hearings before the state Department of Natural Resources the corporate weather modification was ended. Such weather modification could become a proper ballot issue wherever it threatens to shape the weather for all in order to benefit a few.

If such a ballot had been conducted in South Dakota, suggest Drs. Kellogg and Schneider of NCAR, then "a statewide premium could have been levied and a no-fault weather modification insurance policy could have been issued to every citizen who could be affected by the operation." This would have provided some protection for the Rapid City victims. It also would have provided a restraint on the weather modifiers, government or non-government, for the price of such insurance premiums would be determined in part by the strictness of legal, ethical, and scientific controls over their operations.

None of these legal regulations over weather modification can provide complete answers to the issues raised, but they suggest some preliminary steps lawmakers need to consider.

11

Weather, Climate, and War

Deliberate destruction of an enemy's natural environment is nothing new in the history of warfare. When, after three bloody wars, the young Roman Empire conquered Carthage in 146 B.C., the victors sowed salt into Carthaginian fields to destroy their fertility for future generations. In America's Civil War, Union General William T. Sherman marched an army across Georgia and had his soldiers kill every animal, trample every crop, burn every building along their path. "War is hell," he said, and proved it by leaving a sixty-mile-wide zone of devastation in his wake. During the Nuremberg war crimes trial following World War II one Nazi general was convicted of, among other things, the crime of deliberately flooding parts of the Netherlands to punish political dissenters. The list could go on and on, because for thousands of years armies have scorched earth and laid waste to land in trying to achieve victory or revenge.

Such warfare methods in the past had two things in common. They were aimed at the natural ecological base from which an enemy survived and thrived—his farmlands, streams and rivers, forests, and livestock. And they were, almost without exception, tactics rather than strategy. They were used ad hoc, often on the spur of the moment, to achieve quick victory, to shock, to break an enemy's will. They were weapons used in the heat of passion, rather than in cool calculation.

But when, in 1973, Secretary-General Maurice Strong of the United Nations Conference on the Human Environment predicted "that in 10 or 15 years environmental aggression will be a major source of political conflict," the art of war had already changed. A new form of war based on the cold, systematic destruction of an opponent's natural environment had been used in a small nation called Vietnam, and it appeared to be the prototype for all future combat—a frightening form of total war that asks no quarter and offers no mercy, that lays waste to the land with all the sophistication that biological and weather science can provide.

Even today, few Americans understand what their policymakers were trying to do in the Indochina War. The war was fought not only to "prevent a communist takeover" in Southeast Asia, but also to provide the Department of Defense with a laboratory in which it could experiment with new methods of warfare. During the 1950s under President Dwight Eisenhower, America's military policy was simple: Marxist aggression anywhere would be met with "massive retaliation" by U.S. nuclear weapons. By 1960 this doctrine had proved unworkable. How can an atomic sledge hammer reasonably be used against guerrilla movements and the brushfire wars they were stirring up in nations throughout the world? When President John F. Kennedy entered the White House in 1961 he and Defense Secretary Robert McNamara set about creating a more viable military posture for the United States.

Kennedy's new doctrine was called "flexible response," and was dedicated to "defeating the enemy at his own level of aggression." If an enemy was a rag-tag army in the jungles of Indochina, he could be beaten by a small commitment of troops to the jungle. Kennedy ordered 14,000 armed U.S. troops into combat in the long-lived Vietnamese civil war to test "flexible response." Gradually the U.S. military commitment increased, and in 1965, when the new policy seemed to be failing, President Lyndon B. Johnson escalated the conflict with a massive infusion of men, equipment, and new ways of fighting.

In the defense community in Washington, D.C., "flexible response" was thought of as humane. It was, after all, an alternative to the potential horror of nuclear war that had loomed ominously over the Eisenhower years. But as each new military escalation in Vietnam seemed to bring U.S. policy no closer to victory, almost every weapon in the arsenal of democracy was used short of atomic weapons.

America was unprepared for a guerrilla war, just as the British two centuries before had been unprepared for it in their American colonies. Guerrillas attack and then disappear—into the jungles, and into villages where they look like innocent peasants. American military analysts said from the first that an anti-guerrilla war in Vietnam could not be won unless the United States and her allies could put fifteen troops into the field for every guerrilla. Throughout the long conflict U.S., South Vietnamese, and other allied troops never outnumbered the enemy's fighters by more than four to one.

But, said some analysts, victory might come by other means. If the enemy hides in villages, then remove the villages. Vietnamese peasants by the millions were moved to protected compounds, likened by some critics to concentration camps. Those outside these huge strategic hamlets were then regarded as the enemy and often shot on sight.

And if the enemy hid in jungles, the solution was simple: remove the jungles. This tactic never would have been thought of, much less considered, by military planners of the past—in part because earlier technologies were unable to speedily remove whole jungles, and in part because such an idea would have seemed insane. But in 1965 the United States began clearing vast areas of Vietnam, using methods both mechanical and chemical.

Giant bulldozers were set to work uprooting whole tropical forests that had been designated enemy strongholds. The machines were called "Rome Plows," perhaps in honor of Carthage, and the typical type was a 20-ton caterpillar tractor fitted with an 11-foot-wide plow blade. Each tractor carried 14 tons of armor plate. Each plow blade weighed 2½ tons. To attack a forest, typically, thirty such tractors would gang together and move forward in a phalanx, smashing down huge trees and leaving the ground scraped clean where they had passed.

By 1971 the plows had cleared at least three-quarters of a million acres in South Vietnam, and were adding a thousand acres daily to the total.

The land was also cleared by bombing and shelling. In a 1972 *Scientific American* article, Drs. Arthur Westing and E. W. Pfeiffer estimated that between 1965 and 1971 U.S. forces had exploded 26 billion pounds of munitions in Indochina, energy equivalent to 450 Hiroshima nuclear bombs. In the whole of Indochina this averaged to 142 pounds of explosive for each acre of land, 584 pounds for each person. In South Vietnam an overall average of 497 pounds of explosive per acre were used, 1,215 pounds per person. This hurt the land as well as the people. The soil was pockmarked with an estimated 26 million craters. At least 10 percent of the agricultural land of South Vietnam had been abandoned because of this damage.

Up to half the land in Indochina is lateritic. Exposed to open air, it hardens to brick, and as Yale University biologist Dr. Arthur W. Galston writes, "this brick is irreversibly hardened; it can't be made back into soil. . . ." Thus a hole in the ground cannot simply be filled, nor can the land be brought back to fertility easily. "It is doubtful," said Westing and Pfeiffer, "that many of the forests and lands of Indochina can be rehabilitated in the foreseeable future."

Another 5 million acres of Indochina forest and cropland were bombed with chemical defoliants and herbicides designed to kill plants or make trees drop their leaves. This, too, was done "to deny the enemy cover" beneath jungle canopy, and to a lesser degree to deny the enemy food by killing his crops. The problem with this later policy, as Harvard University nutritionist Dr. Jean Mayer warned, is that in a war soldiers get first food priority. However severe the malnutrition in Southeast Asia, the enemy fighters would eat while peasants—especially the old, the young, and the sick—starved. Food destruction was tantamount to war on civilians.

The problem with mass defoliation is that the ecological damage it can do far outweighs its military gains. It cannot be done piecemeal, for an enemy can easily move half a mile to intact trees if leaves began to fall over his base of operations. But mass spraying of defoliant chemicals over wide areas risks destruction of the whole natural web of life in a region, now and for decades or centuries to come. Science cannot certify any large-scale defoliation effort as safe, because as yet our understanding of how living things depend on one another—how trees nurture insects that feed birds that fertilize trees, as one crude example—is too sketchy to enable us to predict what will happen in a forest if we strip the trees naked. In the tropics of Southeast Asia, where soil is thin, and where it has been protected by evergreen trees for millions of years, we must assume defoliation can devastate all traditional life cycles.

Despite these risks and potential problems, the United States carried out a program of mass defoliant and herbicide spraying in Indochina. How much was used is unknown. The highest unofficial estimate for the whole area is 6 million tons. Officially, the U.S. military acknowledged spraying or dumping 50,000 tons of such chemicals in war zones within South Vietnam as of 1970, of which, writes investigative reporter Thomas Whiteside, "20,000 tons have apparently been straight 2, 4, 5-T."

This designation 2, 4, 5-T is shorthand for a type of trichlorophenoxyacetic acid, an active ingredient in America's most popular Vietnam herbicide, "Agent Orange." After many thousands of tons of it had been dropped over Indochina forests, where rain would wash it into the drinking water of farmers and peasants, research in the United States revealed that when fed to pregnant rats 2, 4, 5-T causes terrible mutations and birth defects in every litter born. Such mutations reportedly have increased among humans in Indochina during the past decade, and have appeared in Globe, Arizona, where 2, 4, 5-T was used as a weed-killer near canals carrying water used locally. The chemical does a good job of killing unwanted plants, and presumably of making trees drop their leaves. It also provides what poet Bob Dylan called the "worst fear" that can be hurled at an enemy, "fear to bring children into the world."

The damage from defoliants spreads out in widening circles. For example, wrote Dr. Galston, 2, 4, 5-T devastated up to 40,000 hectares of mangrove trees, many around the Saigon River. "Ecologists have known for a long time that the mangroves lining estuaries furnish one of the most important ecological niches for completion of the life cycle of certain shellfish and migratory fish. . . . In the years ahead the Vietnamese, who do not have overabundant sources of proteins anyhow, are probably going to suffer dietarily because of the deprivation . . . of fish and shellfish." Trees die. Denuded land erodes, dries to brick. The animal and insect life of the forest perishes. Often predators succumb before their prey, which then proliferates. (For example, in the Sahel region of Africa some rains returned in 1974 and 1975, after six years of severe drought. As of January 1976 the nations of Senegal, Mali, and Mauritania in the Sahel were being devastated by a rat plague; in some regions, estimated experts, one rat was alive and eating crops for every square meter of land.) With natural stability and equilibrium destroyed in Indochina, harmful insects and other pests have been on the increase, as have several contagious diseases.

All these changes help influence weather and climate in Indochina. The winds blow differently over bare grasslands than over forests. Stripped, the land is more desertlike. Less rain should fall, but the annual monsoons should bring increasing erosion and flooding. The region should be slightly less humid, but hotter. Crops will be harder to grow. Storms should do more damage than before, and over a wider area for each storm. Vietnam lies in the same belt of latitude as the African Sahel. If the global cooling progresses, it may increasingly suffer from flood and drought, made all the worse by the after-effects of America's experiment with eco-war.

But U.S. environmental tampering was not limited to the land. American bombers modified the rainfall over Indochina, too. Chemicals were dropped into clouds over North Vietnam, and perhaps elsewhere, to give acidity to the rain. This "hydroscopic" seeding, said Pentagon officials, was to foul up radar antennas that directed enemy anti-aircraft missiles. Whether the acid rains also injured living things is unknown, but is a possibility. The question merits study. In some parts of the eastern United States the acidity of rainfall has increased to 1,000 times what it was in 1955, apparently a result of industrial and other pollution.

And the U.S. caused rain. A 1974 congressional investigation found that between 1967 and 1972 the Department of Defense spent $3.6 million each year seeding rainclouds over North Vietnam, South Vietnam, Cambodia, and Laos. During 1967 and 1968, the only years the Pentagon said it seeded clouds over North Vietnam 1,115 seeding units were dropped there.

The declared goal of the seedings, which in at least one instance increased rainfall over enemy territory by 30 percent, was to muddy vital trails and hamper movement of men and supplies.

Pentagon officials said it succeeded, and defended the program as an inexpensive and humane weapon of war. "Raindrops don't kill people, bombs do," said one spokesman. CIA experts privately called the seeding operations a failure.

In 1971, while American aircraft continued cloud seeding over other Southeast Asian countries, North Vietnam suffered the heaviest rainfall since 1945. Floods resulted, and inevitably killed many. The 1945 floods killed more than a million people, most by starvation. The 1971 floods ruined North Vietnam's rice crop and must have caused serious hardship, if not famine. The Department of Defense denied responsibility for the unusual rainfall when questioned about it years later. But before the floods the Pentagon had instituted a policy of bombing dikes used for flood control in North Vietnam.

The military rainmaking was kept highly secret, not coming to public attention until Dr. Daniel Ellsberg released the so-called *Pentagon Papers* in 1970. The papers, a classified history of Defense Department policymaking in Vietnam, referred in Volume Four to weather-modification experiments "successfully" conducted in Laos in 1966, and to a weather-modification program named Operation POP EYE designed to "reduce trafficability along infiltration routes" used by the enemy. Earlier in the volume a list of escalation proposals given to the president by the Joint Chiefs of Staff in 1967 included: "8. Cause interdicting rains in or near Laos."

Syndicated columnist Jack Anderson, in his *Washington Post* column of March 18, 1971, said that U.S. Air Force airplanes had been seeding clouds over the Ho Chi Minh trail, which runs through both Laos and Cambodia, since 1967 under a project code-named "Intermediary-Compatriot."

Pentagon officials said they knew of no ongoing operations by those names, and they spoke the truth. In 1974 testimony before Rhode Island Senator Claiborne Pell's Subcommittee on Oceans and International Environment, Deputy Assistant Secretary of Defense Dennis J. Doolin explained why: Operation POP EYE and Intermediary-Compatriot were the same. Whenever the code names for military rainmaking operations "were uncovered they were changed," said Doolin.

Pell conducted extensive hearings into military weather and

climate modification during 1972 and 1974. A subcommittee chaired by Minnesota Congressman Donald Fraser did the same in the House of Representatives in 1974 and 1975. What emerged was an awesome picture of far-ranging research and experimentation by the Department of Defense into ways environmental tampering could be used as a weapon.

Perhaps the most amazing aspect of the hearings was that they surprised lawmakers. Military people have always been fascinated with weather, especially weather on the days of important battles. This is logical and proper, for weather conditions can determine an army's capabilities, just as a muddy or dry track can decide the winner of a horse race. And the Defense Department had been deeply involved with U.S. weather modifications from its earliest days, flying the airplanes of Project Cirrus and later co-sponsoring Project Stormfury's study of hurricane manipulation. Many states' laws have held that adultery could be proven by as few as two things: motive and opportunity. Military planners had the motive to find weapons potential in any new technology, and they had the opportunity from the first to work with weather modification.

But lawmakers were unaware of how far Defense Department study had gone into military uses of environmental tampering. As expected, the Pentagon had more than half a dozen programs researching control of warm fog, cold fog, cloud dispersal, and, jointly with the Soviets, hail suppression. Less known, the Pentagon had entered the rainmaking business, using its cloud-seeding aircraft to end droughts in the Philippines and the Azores. This had been done, officials said, because both places had important U.S. military bases and rainfall could improve the lives and political well-being of American military personnel. As the hearings brought out, several nations suffering drought and mass starvation in the African Sahel appealed for cloud-seeding help to the U.S. government shortly after the military seeding of the Azores a thousand miles away. (In 1967 the U.S. Navy tried, with little success, to provide cloud-seeding aid to drought-stricken India.) The Sahelian appeals were refused; they were told that private American cloud-seeders could be hired to do the job.

In 1966, Gordon J. F. MacDonald, then a member of the President's Science Advisory Committee and later of the President's Council on Environmental Quality, published a series of ideas for the military use of environmental tampering. "The key to geophysical warfare," he wrote, "is the identification of the environmental instabilities to which the addition of a small amount of energy would release vastly greater amounts of energy." For example, he said, "a controlled hurricane could be used as a weapon to terrorize opponents over substantial parts of the populated world."

MacDonald saw limited military application for rainmaking as a weapon directly, but a good use for it indirectly in stealing an enemy's rain: ". . . a competitor country could be subjected to years of drought. The operation could be concealed by the statistical irregularity of the atmosphere." MacDonald also analyzed how an enemy nation could be destroyed if holes were punched in the atmospheric ozone layer above it, exposing its people to dangerous levels of ultraviolet radiation from the sun. This might be done, he said, by increasing the amount of ultraviolet radiation striking that layer at the specific wavelength of 250 millimicrons. As I published in my syndicated column in 1972, the Defense Department was rushing development of a new type of laser cannon that could produce this precise ultraviolet wavelength; such a cannon, mounted in a stationary-orbit satellite and powered by sunlight, could scan downward and chop a continuous hole in the ozone layer above an enemy target, such as the Port of Haiphong in North Vietnam. This is one of several Pentagon projects aimed at the development of lasers that can direct superintense beams of light on a specific wavelength resonant with molecules in a target; if perfected, such lasers could do more than melt a target—they could shatter its structure, as soundwaves at the right pitch can shatter a drinking glass.

Earthquakes would make good weapons, wrote MacDonald, and so would the tidal waves that undersea earthquakes can produce. If atmospheric electricity can be controlled, lightning storms can be directed against an enemy. And perhaps resonant electrical pulses in the atmosphere could be controlled to damage the delicate biological rhythms or brainwave patterns of people in another country.

The Department of Defense has taken action in all the areas MacDonald described. It has used rainmaking in the Indochina conflict, and studied ways to damage the ozone layer both with lasers and chemical-reagent bombardment. It has studied ways to detect and cause earthquakes through "PRIME ARGUS," a project of the Pentagon's Defense Advanced Research Projects Agency (DARPA). It has studied lightning through the resources of Project Skyfire and hurricane manipulation through its role in Project Stormfury.

And under "Project Sanguine" the Pentagon has studied potential effects on humans of electrical impulses in the atmosphere. In a May 1972 radio broadcast in San Francisco I discussed in detail why the U.S. Navy could not reliably reach its own submarines in a national emergency. Its prime communication with the submarines, I said, depended on only three main transmission sites in the world, and these could readily be sabotaged. The following day Navy communications chief Rear Admiral Samuel Gravely hurriedly called a press conference in San Diego to announce his discovery that the Navy communications system was "at least fifteen years out of date," and that the president might be unable to reach his submarines with codes needed to unlock their missile nuclear warheads in an emergency.

The remedy to this, said the Navy, was Project Sanguine, a plan to bury up to 25,000 square miles of antenna somewhere in the United States. This antenna would be used to transmit many millions of watts energy on very, very low radio frequencies that can penetrate water. What effect would such powerful transmissions have on humans near the antenna? The Navy has tried to build Sanguine in Wisconsin or Texas or Michigan, and in each state has been opposed by environmentalists citing the project's unknown dangers. Nonmilitary research indicates that some kinds of radio transmission can kill bacteria and other living things. Work by Dr. Ross Adey at the University of California, Los Angeles, indicates that the "biological clocks" of living things can be disrupted by transmissions on frequencies not far different from what Project Sanguine would use. Sanguine critic Senator Gaylord Nelson forced the Navy in 1975 to release its research showing that Sanguine-like transmissions can alter human blood chemistry.

As early as 1958 the chief White House advisor on weather modification, Captain Howard T. Orville, declared that the Defense Department was studying "ways to manipulate the charges of earth and sky and so affect the weather" by means of "an electronic beam to ionize or de-ionize the atmosphere over a given area." Perhaps then-Senator Lyndon B. Johnson had this in mind when, in 1957, he told the assembled House and Senate: "From space one could control the Earth's weather, cause drought and floods, change tides and raise the level of the seas, make temperate climates frigid." (Captain Orville also discussed ongoing U.S. Air Force experiments with "sodium vapor, ejected from jet planes, to intercept solar radiation" over enemy countries and ruin their weather.) But as Stanford University scientist Robert Helliwell revealed in 1975, we are already inadvertently altering the ionosphere above our heads with the Very Low Frequency radio energy emitted by all electrical power lines, whose VLF output is on wavelengths much like those planned for Project Sanguine and already used by military broadcasting facilities in Antarctica and elsewhere. While working in Antarctica Helliwell discovered that these VLF signals can be multiplied up to 1000 times by a coherent interaction with particles in our planet's radiation belts above the Earth; the radio signals cause an "electron rain" in the ionosphere, and this inevitably alters weather patterns in the lower atmosphere. The Defense Department is already studying this phenomenon.

MacDonald proposed one other environmental weapon: climate modification. A "landlocked equatorial country," he wrote, might see advantages in raising world sea levels by melting the polar caps, or in using technology to hasten global cooling and the dawn of an Ice Age. He assumed, wrongly, that an Ice Age would improve climate in tropical regions; it would more likely cause drought.

The Department of Defense has been fascinated by climate modification since the early 1950s, when U.S. military intelligence learned of a Soviet plan to ruin climate in the United States. The plan was to build a jetty 50 miles or more long out from near the eastern tip of Siberia. The jetty would contain several atomic-powered pumping stations that would push cold Arctic waters down through the Bering Strait between Siberia and Alaska. This would cool the Bering Sea and inject increasing amounts of icy water into the ocean current that flows down along the west coast of Canada and the United States. The result would be colder, more stormy weather throughout North America and enormous losses to the American economy in agriculture, work days, and storm damage. Military experts already knew about the beginning of the global cooling trend, and they had to wonder whether the Soviets were to blame for it. In retrospect, this seems illogical: the Soviet Union has been hurt worse by the cooling than most other nations, and by its plan hoped to set up a new ocean current drawing warm Gulf Stream waters into the Arctic. But Pentagon planners worried and started studies of ways to change climate.

Defense Department climate research coalesced in the early 1970s in a DARPA project called "Nile Blue." In fiscal 1972 the project had a budget of $2.587 million to develop and monitor computer "models" of changes in world climate. For this purpose it was provided the most sophisticated computer yet developed, the "ganged" ILLIAC IV system, which alone could handle the mass of data and subtle changes that make weather prediction so difficult.

But about that time political controversy was growing in the wake of revelations about Pentagon rainmaking in Vietnam. Should the military be working with weather and climate modification at all? asked critics and a few lawmakers. Its budget threatened, Nile Blue changed its name to "Climate Dynamics" and DARPA, under whose control it remained, announced that henceforth the program was completely open and unsinister. In 1973, in a seemingly unrelated event, Dr. W. Lawrence Gates announced that the Ames Research Center in California had just acquired a new ILLIAC IV "ganged" computer from DARPA, which he would be using for world climate modeling. Dr. Gates is head of climate research at the RAND Corporation in Santa Monica, California. The RAND Corporation was established in 1948, at the beginning of the "Cold War" between the United States and Soviet Union, to serve as a civilian intelligence arm of the Department of Defense. Now diversified, it still does classified military studies for the Pentagon. As a private organization, the RAND Corporation is not subject to the same degree of congressional scrutiny as the Department of Defense.

In 1972 congressional hearings before Senator Pell's subcommittee, the Defense Department said it was conducting no classified *weather* research. Climate research was unmentioned. It defended Nile Blue as giving military planners a needed "capability to predict the climatic effects of foreign actions and to detect modifications which may be in progress. With a scientifically credible detection capability, world opinion and the instruments of national power may be mobilized to reverse actions damaging to the national interest.''

''The Soviet Union,'' said the Defense Department in response to questioning by New Jersey Senator Clifford Case, ''has invested considerable effort and resources in developing a well-organized and extensive program in climate modification research. The Director of the Soviet Hydrometeorological Service has declared that active modification of climate is an objective of this research. A number of specific projects have been proposed to alleviate the harsh Russian climate with attendant benefits to agriculture, navigation, and resource exploitation. These include removal of the Arctic pack ice, damming of the Bering Strait, and diversion of Siberian rivers.

''These programs,'' the Defense Department statement continued, ''clearly might affect the climate of other parts of the world, including the United States and its allies. Even marginal changes in temperature and rainfall could drastically damage agriculture, shipping, and indeed the entire economy. Military operations would also be impacted if the boundaries of pack ice, the ice-free seasons of naval bases, the frequency of obscuring clouds, etc., were altered. Thus climatic changes are clearly potentially grave threats to national security, and have consequent implications for military planning.''

The Pentagon's position was that Nile Blue had too much military significance to be transferred to civilian control. It has never explained how, under military control, the program can serve as a ''scientifically credible'' instrument able to sway ''world opinion.''

Could such a computer climate-modeling program as Nile Blue have aggressive military potential? Certainly. Its findings can provide military planners with knowledge of those key ''environmental instabilities'' Dr. MacDonald said were essential to the successful use of environmental warfare. As Robert M. White, Administrator of the National Oceanic and Atmospheric Administration, has said, ''It is not possible to draw clear distinctions between research and technological development on weather modification for hostile and nonhostile purposes." "Obviously," said Maryland Congressman Gilbert Gude, "such knowledge can be used for offensive military purposes."

In 1975 Senator Pell and Representatives Gude and Fraser, the three leading legislative critics of American military research into weather and climate modification, sent a letter to President Gerald Ford urging increased government support for peaceful uses of such modification.* They also urged that *all* such research and operations, military and non-military, be placed under the control of an oversight civilian agency answerable to the president and to Congress.

Officials admit that Pentagon climate modeling has studied ways and means and probable results of melting the polar ice caps, and the possible consequences of Soviet schemes to melt the Arctic ice pack. They have also studied how to make and direct tornados and hurricanes, and how to destabilize weather in the Soviet Union, China, and Cuba, which would ruin their harvests and thereby strengthen the U.S. "food weapon."

Such research is controversial because it seems contrary to American ethics of warfare. Environmental warfare, especially involving weather and climate modification, can be used against an enemy secretly, as the U.S. did over Indochina for several years with its rainmaking operations. One limitation of its use is imprecision: a storm will hit civilians as well as enemy soldiers, and it may cause unforeseen secondary effects that will harm whole populations. American declared policy has long been to avoid waging wars on whole peoples, to renounce weapons that injure innocent civilians. This was one reason President Richard Nixon ordered destruction of all American germ weapons, for their use, like that of weather or climate modification, could not be limited to precise targets. The United States has never explicitly renounced environmental modification as a tool of policy.

During a summit meeting between President Nixon and Soviet Premier Leonid Brezhnev on July 3, 1974, the nations agreed to conduct discussions toward a ban on environmental warfare. Before the first of these discussions, set for Moscow in November, got under way, the Soviet Union introduced a resolution before the

*See Appendix Five.

United Nations General Assembly to ban environmental warfare. When revised, the resolution was passed by the body 102 votes to none. The United States and half a dozen other nations abstained from the vote. Senator Pell suspected that the president felt miffed by the surprise Soviet action, a move that made it appear that the Soviet Union and not the United States had taken the lead in trying to ban environmental modification. In fact, the Soviet resolution was similar to one passed by the North Atlantic Assembly in November 1972 and to another authored by Senator Pell and passed by an 82 to 10 vote by the United States Senate in July 1973.*

All these resolutions present us with problems and ironies. Because weather and climatic war can be secret, how shall we detect them without international inspection of some sort? The Soviets have always balked at such inspection. What constitutes military versus peaceful use of environmental modification technologies? Is the difference one of effect or intent? The Soviet resolution before the United Nations, for example, forbade "modification of the natural state of the rivers, lakes, swamps and other aqueous elements of the land" if this produced "harmful consequences." But the Soviets are already in the process of reversing the flow of Pechora River, whose waters naturally flow into the Arctic Ocean. Reversing such rivers, experts warn, may warm the Arctic by reducing its total fresh-water input, and hence diminish the ice pack. The Soviets have also dammed several rivers to create inland seas. Such actions will have a climatic impact far beyond the borders of the U.S.S.R.

Two weeks after the first U.S.–Soviet discussions on banning environmental warfare, President Ford and Soviet leaders met in Vladivostok, Siberia. One of the items secretly discussed during this summit meeting was the state of joint Soviet–American research in the Arctic and the Bering Sea. This research had been undertaken to evaluate the cooling trend in Earth's climate, consider what threats it posed to both nations, and discuss whether weather and climate modification technologies should be tried in dealing with it. One idea long considered by the Soviets is damming the Bering Strait between Siberia and Alaska. This possibil-

*See Appendices Two–Four.

ity was mentioned at the Vladivostok meeting. Information gathered from the joint U.S.–Soviet research effort in this region would be used, it was agreed, in analyzing any such plan.

Any such attempt at global climate modification could be seen as a hostile act by a nation whose weather worsens—and because the global cooling trend has already made weather less stable everywhere, many critics will call for the prohibition of all major climate-modification programs.

Discussions between U.S. and Soviet negotiators resumed in Washington, D.C., on February 24, 1975. On August 21, 1975, the two nations presented their jointly produced draft treaty banning environmental modification as a weapon of war to the thirty-one-nation Geneva Disarmament Conference.* It prohibits signatory states from engaging in "military or other hostile use of environmental modification techniques having widespread, long-lasting or severe effects as the means of destruction, damage or injury to another party state." Would such a provision ban Soviet river modification? Perhaps not, for it is neither military nor hostile. Would it forbid U.S. cloud seeding in Vietnam? Perhaps not, for the effects of such seeding admitted to by the United States were neither widespread, long-lasting, nor severe; as one military analyst said in defense of the seeding, "People in Southeast Asia are used to heavy rains."

Climate modification would most likely be used as a form of economic warfare, to destroy crops or hamper transportation or production. In the case of Soviet river modification, such damage might come to China purely as a by-product of efforts by the Soviets to improve their own climate. And the draft treaty is explicit: nothing in it shall be taken to prohibit or limit peaceful methods of environmental modification a nation uses to help its own people. Also nothing in it provides for international inspection or monitoring, without which treaty violations would be difficult to detect or determine. Indeed, if the global cooling trend continues and world food production suffers, people will begin demanding that weather and climate modification be tried to offset damage done by natural forces. Few will want these methods of warfare banned. Many will want them to be directed against the cooling itself.

*See Appendix 4.

PART III

Options
in a Changing
Climate

Alternative I

Change Ourselves

When climate changed in ages past, our ancestors had only one choice: to survive, they adapted themselves to new conditions. Often this meant migration to lands where climate was better. Always it meant challenge, a test of human intelligence, courage, and flexibility. Today's changing climate and the threat it poses of a possible new Ice Age are little different from what earlier peoples faced. Humankind can endure once again if it can adapt to such changes.

We have virtually lost one option in recent centuries. The world's land is now occupied by governments with awesome armaments. All countries now restrict immigration, even those, like Australia, that have vast unpopulated territory. During the past 100 years, wars and famines have prompted more human migration than ever before, but such movements are now generally foreclosed. As climate becomes worse throughout the world, we cannot expect to escape it by migrating as people once did. We must prepare to stand and fight—fight drought in lands near the equator, and perhaps fight a new "cold war," a war against expanding cold, in all the lands above 35 degrees North Latitude.

Here we shall consider a few of the possible ways in which humankind might adapt to the cooling of our planet. Many of these ideas involve concepts and methods people have used for thousands of years to survive in hostile climates. They suggest how we can change ourselves. Whether we can change weather and climate will be dealt with elsewhere.

Human civilization derives from only a few revolutions in the way our species lives. We moved from trees into caves. We changed from hunters into farmers. We extended our power with simple tools, then machines, then electronic technologies. But when we think about these advances, looking for the ways they helped our ancestors survive, it becomes immediately obvious that these revolutions were never completed or consolidated. Perhaps by carrying on the ideas humankind developed long ago we can gain the food and energy and shelter to survive all the Ice Ages that sooner or later will come. In the roots of our past we may find our future.

World Under Glass

"Cave man" was the first human being to create an artificial micro-climate. Inside the closed air space of a cave he would light a fire and huddle his family around it for warmth. This heated the air inside the cave, where springtime temperatures could be maintained during coldest winter. When snows came, the cave dwellers brought firewood and dried food inside. Whatever plants they harvested were left to die outside in the harsh weather.

In these regards we have advanced very little over ancient cave people. We build our own caves, and like the caveman we use costly fuel to create artificial climates inside them. In the United States the heating and air conditioning of buildings consumes more energy than any other single activity—which explains why the president's first request during the 1973 Arab oil embargo was that citizens readjust thermostats where they lived, worked, and drove to more nearly "natural" temperatures: hotter in summer, cooler in winter. This saved a great deal of energy. Like cave dwellers, too, when winter comes we move indoors and leave the plants in our fields outside to die. We live on food preserved since the growing season or transported at great expense from lands where the climate is better.

Suppose we carry on the revolution the caveman began and bring our plants inside. For centuries peoples in far northern latitudes have used greenhouses to grow tropical plants or out-of-season vegetables. The principle is simple: sunlight passes through the glass of the house, is converted to heat when it strikes things inside, and is trapped because the windows of the greenhouse prevent its escape. Until recently the cost of building large greenhouses was prohibitive, but as the price of food rises this has been changing. New building materials have brought cost down as well. In 1974 Dr. Theodore Taylor, head of the International Research & Technology Corporation, estimated that "grains, fruits, and vegetables sufficient for providing a balanced diet for more than 200 people from each hectare of greenhouse structures can be realized at initial capital costs as low as $50 per consumer." Thus greenhouse agriculture is economically feasible today in developed nations.

The advantages of greenhouse agriculture are many. It would protect crops from cold and storm, thus eliminating frost and hail damage. It would extend growing seasons in northern countries, and make agriculture possible where it never before was. In dry or hot lands it would hold in water that otherwise would be lost to evaporation. Insects and other pests could be kept out, or at least more easily controlled, in greenhouses. Less pesticide would be needed, and honeybees could live and pollinate securely within the structures. Protected from wind and rain and storm, the world inside the greenhouse would suffer little erosion; soil could be maintained with a minimum of artificial energy subsidies. Green plants breathe carbon dioxide, not oxygen; inside the greenhouse the stifling atmosphere of the natural world could be replaced with one richer in CO_2, and the plants will respond by thriving and increasing food yields.

With greenhouses we can grow food on all those lands now too marginal for reliable production. Global food harvests would grow steadily, unaffected by bad weather or cooling climate. Taylor and Charles C. Humpstone estimate that "seven billion people could be fed with the annual output from 470 million acres under greenhouses, or about one-seventeenth of the total land now used for agriculture." And as Taylor notes, greenhouse agriculture "can be either capital- or labor-intensive"; i.e., less developed countries can use human labor to farm the crops, and developed nations can use costly machines and chemicals. Thus such agriculture can be employed cheaply by any nation. Its value will grow as unstable climate disrupts global weather more and more.

If we can put our croplands under glass (or plastic, in the case of many new greenhouses), why not our cities? Inventor R. Buckminster Fuller thinks this is both possible and practical. A dome, he writes, could be built over mid-Manhattan, "reaching from the Hudson River to the East River, at 42nd Street, and from 22nd Street to 64th Street . . . a hemisphere two miles in diameter and one mile high at its center. The peak of the Empire State Building's television tower would reach only one-third of the distance from the ground to the dome. The dome would reduce energy losses in winter heating and summer cooling to one eighty-fifth of the present cost. Snow plows would not be needed. Money saved in ten years would pay for the dome."

Is Fuller visionary? In any event he is a genius, inventor of the geodesic come and a host of other innovations. Fuller wants many cities put under giant domes, some for protection, others for preservation as museums showing how humankind once lived. Future domes, he predicts, can rise so high above cities as to seem invisible. Such domes will act like "a controlled cloud, bringing shadow when shadow is desirable, bringing sun when sun is desirable" by moving adjustable shades and polarized lenses. They will keep out rain, storms, snow, and fumes from nearby industry. The stable temperature inside will be semi-tropical twelve months a year. For the dome dwellers—primitive "dome people," descendants may call them 10,000 years hence— "windows may be open year-round. Gardens will bloom year-round. There'll be a dust-free atmosphere. The domed city will indeed be nothing to sneeze at."

Both the United States and the Soviet Union maintain dome cities in the Earth's polar regions. The U.S. cities make use of geodesic domes. Prompted both by the global cooling and the desire to have adequate bomb shelters in case of war, the Soviet Union has rushed development of a subway-connected "underground city" beneath Moscow. China has done likewise in several cities. In Moscow the subway is a subterranean community of shops, recreation areas, and other facilities having many of the advantages of Fuller's domed city. If we expect ever to settle large populations near the poles, or in deserts, or on ocean bottoms, or in colonies on alien planets, development of dome-city technologies seems wise.

At present domed cities are experimental. They will not be built in temperate climates, writes Fuller, until the technology is proven and costs are reduced—unless "environmental and other emergencies make immediate doming imperative." If the cooling continues for a few more decades, Europe, Canada, and the United States will find their climates no longer temperate—and that emergency may prompt city domes and a rush to put crops in greenhouses.

Green Revolution III

Despite thousands of years of refinement, the ways we get food are primitive. We catch fish, for example, as cave dwellers caught game—by tracking them down, by hunting. The human agricultural revolution needs to be extended into aquaculture.

Fish can be raised like cattle in closed ponds with much more efficiency than wild fish can be caught. Carp are raised thus in Israel, China, and central Europe, and Israeli researchers now routinely produce 12,500 pounds of the protein-rich fish per hectare of pond each year. Scots and Norwegians fence off ocean bays to breed salmon. In America pond-farming is used to produce catfish and trout. At present the technology is costly, but with study this can be improved and extended to shrimp, plankton, and other edible creatures.

More efficient crops can be grown at sea. Off the California coast the U.S. Navy runs an experimental seaweed farm; the goal is to harvest seaweed for fuel, but its potential as food is also under study. Planktonic small sea animals, such as krill, are already being harvested in Antarctic waters by the Soviet Union, which processes them into a protein paste.

The Soviets are researching ways to farm such tiny marine creatures. The thin surface layer of the oceans, which scientists call the Neuosphere, teems with microscopic life. American and Soviet experts say that if pipes were sunk into deep ocean waters, the submarine pressure would make it easy to bring this deep water up through the pipe. The deep water, rich in nitrogen and other nutrients, would fertilize surface waters and make continuous harvesting of Neustons possible. However, this work may be thwarted by other events. Every year oil tankers spill more petroleum on ocean surfaces. Eventually this may limit life on the surface of the seas, and that would disrupt ecologic food chains in ways we do not yet understand.

But some micro-organisms "eat" oil. In 1975, General Electric

Company announced it had developed one such tiny creature that could be used to devour tanker oil spills at sea. (They did not discuss its weapon potential: imagine unleashing a colony of the wee beasts in an oil company's storage tanks, where they could feed and multiply.) British and American scientists have for more than a decade developed edible micro-organisms that can be grown in petroleum or other inedible solutions. These are among the many single-cell proteins (SCP) being perfected as ways to produce maximum nutrition value for minimum cost. Given the expensive equipment needed to maintain precise temperature and humidity for growing them, such SCPs can already be produced in quantity anywhere, regardless of external climate. The problem with them, as with most experimental foods, is lack of acceptance by consumers, who prefer traditional tastes and textures. During the 1950s the U.S. State Department tried valiantly to feed more people in India with new foods; it failed, and famines followed. Many people, surprisingly, would rather starve slowly than accept a revolution in what they eat. (We could feed the whole world today if people everywhere were willing to eat weeds, animal blood, and insects, and foresake meat, each pound of which requires at least seven pounds of grain to produce. In the United States meat is gradually being replaced by "textured vegetable protein," now commonly eaten in "TV dinners" and in "meat extenders" which homemakers add to ground beef—but which ironically cost them more per pound than hamburger.)

Lack of acceptance was one problem with new grains of the "Green Revolution." In Mexico, for example, Dr. Norman Borlaug introduced a new "opaque" corn that grew better than the old and yielded significantly more protein per kernel; acceptance was poor because meal from the new corn had a consistency different from the old, and natives found that their tortillas tasted strange.

In the old revolution scientists concentrated on only two things: increasing food yields and enhancing food value in grains. Today "Green Revolution II" is underway. Scientists now work to produce new foods that will be acceptable to native populations and that will have improved resistance to pests and changes in climate. Grains of the old revolution are often re-crossed genetically with hardy traditional grains to make new hybrids that are more fruitful than the old but stronger and more adaptable than the

Green Revolution I grains. The process is slow: a new hybrid may require fifteen years to perfect. But work is underway, and "genetic banks" are being created to guard against loss of traditional grain traits for future re-crossing if the new hybrids fall prey to disease and other problems. One pure cross between wheat and hardy rye has been perfected: a new grain called Triticale. And in the tropics newly-developed hybrid palms are feeding thousands of people.

Like its predecessor, Green Revolution II is basically an old-fashioned enterprise. A third Green Revolution is emerging, based on the new science of genetic fusing to create whole new species of plants, rather than merely variations on the old. Scientists in this field already talk of perfecting a potato-tomato plant that can bear fruit above and below ground simultaneously, for example. Cells of at least a dozen plants have already been fused. Corn and soybeans have been merged, as have chick peas and sweet clover. And, perhaps strangest of all, a unified plant has been created out of the genetic material of soybeans and Douglas fir; such a creation could be harvested for beans for a few years, then cut for lumber.

In the United States alone 213,000 barrels of oil are used every day to make needed chemical fertilizer to restore lost nitrogen to croplands. With cell fusing and other techniques, scientists are studying ways to let plants produce their own nitrogen, thus saving this petroleum. A few plants, such as clover, do this naturally, and for centuries these were planted in fields and plowed under every few years to maintain fertility in the soil. Such "crop rotation" has largely been abandoned with the advent of chemical fertilizers. In Great Britain it has stopped totally. But the rising prices of oil and fertilizer may bring it back. Clover and its kin produce a root structure in which live certain bacteria, able to "fix" nitrogen in the air and make it usable by the plants. Scientists have been developing ways to modify the roots of corn and other grains so they too can "fix" nitrogen.

Other research has revealed that the chemical process inside green plants uses photosynthesis to merge carbon with both oxygen and carbon dioxide. When carbon merges with carbon dioxide, energy is produced in the plant; when it merges with oxygen, the product is wasted, breathed out by the plant in a process called "photorespiration," which accounts for half the effort of some plants. Chemicals and genetic fusing are being tried to reduce such photorespiration, which would let plants grow faster, stronger, and more efficiently.

Green plants probably evolved on our planet when the atmosphere had much more carbon dioxide than it does today, and they thus evolved in climates more tropical than we now know. Corn is scarcely growable in the Soviet Union, but even in the United States, say experts, the heart of the corn belt is 300 miles farther from the equator than it should be—an error caused by habit and tradition. Experiments are underway to improve photosynthesis in plants, which experts say can be increased by more than 100 percent; if successful, this would let plants grow more abundantly in warm climates, and make growth possible where climate is cold. Work is underway, too, to increase carbon dioxide available to growing plants. Dry ice has been sown into fields. Microcapsules of the gas have been plowed into soil. CO_2 has been mixed into irrigation water and also injected directly into soil. All these experimental methods have increased plant growth and yield by giving green beings of our planet air more like what their ancestors knew.

Humankind domesticated animals for food long ago, but only recently have we begun to apply advanced technology to the enterprise. Some methods have proved dangerous; among the chemicals used to fatten cattle, diethylstilbestrol (DES) has proved to be cancer-causing and been discarded in most countries. In science today the goal is not fatter cattle, but more: cows, which normally have one calf per year, soon may bear two if experiments at the University of California succeed. An extra fertilized egg can be implanted in a cow's Y-shaped uterus, and this would double cattle production. This has been done experimentally with cows and pigs already.

As genetic manipulation is perfected, science will be able to clone single cells of useful animals. Cloning is a stimulus that can make a single cell, whether an egg or not, use its genetic code to grow a new, complete organism. The cell of a cow could be cloned, then implanted in the womb of a rabbit for incubation, then later transferred to the womb of another cow. Scientists have already cloned the large eggs of frogs successfully, and by some reports have transplanted ''test-tube babies'' from one human mother to another. It may soon be possible, and socially acceptable, to eat tissue culture—not a chicken, for example, but chicken flesh grown from an endlessly expanding mass of cells nurtured in a vat of chemicals. This process would enhance food-producing efficiency by eliminating the middlemen, the chickens and the eggs.

Perhaps, just as synthetic microbes and plants have been created by genetic fusion, animals can be created. United Nations researchers are now studying the oryx, an antelope of the African Sahara that can survive on one-tenth the water cattle require. A cross-breed of cow and oryx might produce cattle able to thrive in the driest brushlands of the planet. Steers have already been successfully crossed with American bison to produce the highly meat-productive ''beefalo.'' The possibilities of this technology are limited only by human imagination, good or bad. It certainly could be used to improve climatic adaptability in creatures.

According to researchers at the University of Michigan, humans would benefit greatly from lower body temperatures. We all could live 100 years or more if our body furnaces ran 2½ degrees C. cooler, and perhaps 200 years if our body temperature could be reduced by 5½ degrees C. Would this help us adapt to cooler climate? Perhaps, if only because maintaining a lower body temperature would require less food intake, and thus more people could survive on less food.

New Sources of Energy

Just as the global cooling has threatened our food supply, so too it puts new demands on our energy supply, more and more of which must be allocated for heating, air conditioning, fertilizer, and other uses influenced by climate. Many potential sources of new energy remain virtually untapped.

Atomic fission reactors will be producing a large share of the world's electricity by year 2000. To do this they will use increasingly-scarce nuclear fuel, produce dangerous radioactive wastes, and contribute to a stockpile of plutonium that, if stolen by terrorists or diverted by governments, could readily be turned into atomic weapons. Research is now close to perfecting the technology for a fusion reactor, a "controlled H-bomb," that would generate vastly more energy than fission reactors. It would use "heavy water," which can be obtained in almost unlimited supply on this planet. It need not produce dangerous radioactive by-products in the quantities a fission reactor now does. It would create a "fusion torch," a surplus stream of heat which, at more than one-half million degrees Kelvin, could reduce any kind of waste and pollution back into its basic atomic elements—thereby eliminating the human garbage that threatens to engulf us.

Fusion power has drawbacks. With it, vast thermal pollution would be inevitable, and this could kill wildlife and influence world climate. Preliminary plans for such a reactor would put it on the Grand Bank off Newfoundland. If, owing to malfunction or sabotage, it exploded, the ensuing tidal wave would swamp Boston, New York, Miami, and the Netherlands. This, plus subsequent radioactivity, could kill millions and have unpredictable climatic impact. Prolonged safe operation of such a reactor would warm the Gulf Stream/North Atlantic Drift, which would temporarily improve weather in Europe but soon would melt ice in the Arctic Ocean.

The sun delivers great amounts of energy to our planet constantly, 1 percent of which would supply all human needs for thousands of years. The U.S. government in 1975 committed 25 percent of its energy research budget to exploring ways to tap solar energy. Especially in the tropics, where more than half of all this sunshine reaches Earth, solar collectors on housetops are good investments now. But in northern Canada and the Soviet Union some seasons receive so little sunlight that the technology seems of limited value, even with heliotropic, sun-tracking receptors.

But we need not confine ourselves to sunlight on Earth. Studies by the Boeing Company and Arthur D. Little, Inc., indicate that we can feasibly create giant solar collectors in outer space that will gather solar energy as light, convert it to radio frequencies, and beam it to Earth as microwaves, using highly directional antennas, with losses of between 10 and 25 percent. Research on Earth has already successfully demonstrated that large amounts of power can be transmitted this way. The 1968 proposal of Peter Glaser, an Arthur D. Little scientist, called for two giant square solar collectors, mounted in space by a joining microwave transmitter. This could be put in space as soon as the U.S. space shuttle becomes operational.

With a few hundred such stations in space, all power plants could be closed down on Earth, and with them would end much pollution, thermal and otherwise. The idea is economical and attractive. One drawback: microwaves kill. Military personnel have died from walking in front of radar antennas. Nobody fully understands why microwaves are dangerous, but apparently they can damage living cells. In ovens, all of which are shielded and require locks, even low-energy microwaves can quickly penetrate and cook food. What would happen if billions of watts of microwave energy began bombarding the Earth? Glaser says the danger is small, but that living tissue could conceivably be harmed. A person struck by these beams would receive nine times what the Public Health Service calls a safe dose of microwaves. He also says the energy would be too low to ionize the upper atmosphere, but because our atmosphere exists in an electro-magnetic field we must assume it could influence winds, weather, and climate. Power plants on Earth already disrupt the ionosphere.

The sun and moon already move wind and water on our planet, and we already derive some energy from this movement. Hydroelectric dams merely catch falling water that was lifted by evaporation from the seas and convert its motion into electricity. But the wind moves, ocean currents flow, and tides rise and fall too. These can be tapped as clean, safe, nonpolluting sources of energy.

We have all seen and read about windmills. Don Quixote tilted at them. Highly efficient models of them have recently been created. Their limitation is that they do not work when the wind is still, but in the U.S. midwest, wind averages better than 10 miles per hour.

Tides have been used to run a power station at the mouth of the River Rance in France for more than a decade. Recently the Soviet Union set up a similar station on Kislaya Bay in the Barents Sea. When the tide is high, the station closes its gates and traps the water, which it then slowly releases through turbines during low tide. The United States and Canada are considering construction of a tidal power station in Passamaquoddy Bay between Maine and New Brunswick, where President Franklin D. Roosevelt lived on Campobello Island and talked of ways to harness the tide.

Such stations are expensive to build (Rance cost $100 million) but cheap to operate, for they derive their power from the moon and its pull on the oceans. Technology for tidal stations is as yet crude. Only a few places in the world have a big enough difference between high and low tide to make such stations effective large-scale producers of power, given present technology.

Just as tides can be captured at their height and used to produce electricity as they fall, so too can waves be caught and used hydroelectrically, say some inventors. Small workable models of wave-powered generators have been built. Their effectiveness varies with the height of ocean waves, at different times and on different oceans. To produce large amounts of power, rows of these generators would have to extend across thousands of square miles of oceans, and this might impair shipping routes.

Ocean currents are a steady source of power, and no current has stirred more inventor ideas than the Gulf Stream, which flows 15 miles off the Miami shore with a volume of water fifty times greater than all the world's rivers, moving day and night at a speed of four knots. A 1974 meeting of scientists and engineers concluded that 2000 megawatts of electricity could be safely generated by using the Gulf Stream off Miami as a river to drive underwater "windmills" or other turbine devices. To take more than that, they decided, would weaken the stream, which might cool the climate in Europe.

Another means of tapping the Gulf Stream and other currents has been proposed by Clarence Zener and Abrahim Lavi of Carnegie-Mellon University. It involves putting a generator at the point where a warm current verges on colder ocean. Within a closed system a chemical would be pumped that was liquid at the lower ocean temperature, but became a gas at the warmer current temperature. This working fluid could be propane, ammonia, or fluorocarbon, all of which expand when they turn to gas. This expansion would supply the pressure to turn a turbine and create electricity. This same principle could be used in a giant generator that would extend from warm ocean surfaces to cold layers near the bottom. Wherever a strong difference of temperature exists, able to vaporize and recondense a fluid, electricity can be generated. Dr. Lavi says with an initial investment of $20 million a one-megawatt ocean thermal plant could be operational within two years, perhaps off an island like Hawaii where government agencies are now considering such a facility.

To create power, these plants would remove some heat from tropical waters, and this would reduce evaporation from the oceans. According to Dr. Zener, ". . . by what amount would we have to reduce the rate of evaporation in the tropics to supply all the energy now being used in the world? . . . three percent. I don't know whether the tropical countries would complain about being a little less wet and a little cooler." The world might complain if this hastened climatic change, or caused drought in moonsoon-belt nations dependent on tropical evaporation for rainfall.

Geothermal power plants, using heat from deep in the Earth's crust, could provide cheap energy. The technology is far from perfected, however, and only a few volcanic regions on the planet have easily-tapped geothermal heat—Iceland, California, and New Zealand, for example, already have such plants. A study of the New Zealand plant suggests that thermal pollution is a problem with geothermal energy; the vast amounts of heat they release into the environment can disrupt ecological systems and potentially disrupt regional climate.

Ideas for new energy sources are many. Among those beginning to win attention: methane gas extracted from garbage; alcohol distilled from plants as a gasoline substitute; heat derived from burning fast-growing bushes that could be grown in wastelands.

What most of the ideas for increasing our energy supply have in common is that they are nearly all ignored by government and private groups alike. Most are still on the drawing board. A few are being tried in experimental models. The old technologies of fission reactors and fossil fuels are well established as habits in the industrialized nations and as vested political interests; changing them will prove more difficult than changing a nation's eating habits. The U.S. commitment to solar power is encouraging, but it is a small first step on a long and urgent journey.

As the United States prepared to celebrate its 200th birthday, and as President Gerald Ford talked of achieving energy independence for the country by 1985, a U.S. Library of Congress study released in December 1975 revealed that American dependence on imported oil will increase sharply by 1980. By 1977, the study showed, the U.S. will be importing 9 million barrels of oil a day, as compared with 6 million barrels in 1975. Owing to anticipated cutbacks in Venezuelan and Canadian exports, moreover, 80 percent of total U.S. imports in 1977 and thereafter will come from Arab nations. Domestic oil production in the U.S. will continue to decline by 8 percent per year from 1975 onwards. "Energy" the study says, "may well become the 'Achilles heel' of U.S. foreign policy the same way as agricultural shortages are for the Soviet Union."

Conservation and Control

Human civilization has advanced, according to Buckminster Fuller, when people learned to "do more with less," to use intelligence and imagination to invent better, easier, cheaper uses of resources. In addition to tapping new resources, we need to improve the ways we use what we already have available.

Take lawns, for example. In the United States millions of hectares of arable land that could produce food are used for decorative grass and shrubs. If the chemical fertilizers Americans use on their lawns, golf courses, and cemeteries were used instead on Asian croplands, writes James Grant, president of the Overseas Development Council, they would grow enough extra food each year to feed all the people of India and China.

Americans regularly mow their rich grass lawns, then throw the clippings away. If these clippings were put into mulch piles and refined by the heat of their own decay, the resulting humus could restore those lawns' fertility year after year virtually without use of costly chemicals . . . or could enrich soil around most homes so that the average American family that chose to replace lawns with gardens could grow 25 percent or more of the food they need each year.

Note this word "need." It is readily confused with the word "want" in industrial societies, where the dominant value is consumption rather than conservation. Ultimately the pressures of unstable global climate, expanding world population, and diminishing natural resources will destroy societies unable to distinguish needs from wants and adjust their ways of life accordingly.

The course of human history changed, said Fuller, when people departed from land to sail in ships, to fly in the air, and to rocket into the heavens. Living in these new fluid dimensions required people to take only what they needed, and to compress that into the most efficient, compact form. Lives lived on land were comparatively simple, for the land was always dependable, always stable underfoot, and life never threatened one with sinking, or falling from the sky, or becoming helpless between Earth and the moon. Life on land was inert, and, said Fuller, "inertia, unchallenged, promotes careless philosophy." The land only becomes fluid under rare circumstances, as when the rains fail or climate becomes unstable. At such times history forces efficiency, frugality, and hard thinking on the people of the land. This is happening today, and it demands a rethinking of our fundamental values, a careful look at what we are wasting that could be wisely used.

Above the oil fields of Nigeria, Venezuela, and the Persian Gulf, thousands of tall smokestacks burn day and night like perpetual candles. They are flaring off natural gas, which is pumped up from the ground with oil. In these three regions enough natural gas is burned meaninglessly to feed all the less developed nations of the world, if it were processed into fertilizer. This would be the most important gift such oil-producing nations could give to hungry countries, but thus far they have been too lazy to do so.

In many U.S. croplands, where such chemical fertilizer has been used for decades, a natural limit has been reached. Grains give diminishing yields each year for the same amount of fertilizer, but using more of the chemicals would risk nitrogen poisoning of ground-water supplies. Insect damage is increasing because another risk ceiling has been reached in the use of pesticides, also derived from costly oil; the poisons have infiltrated water supplies, have killed more than 12 percent of beneficial honeybees on the planet, and have been proven in some cases to cause cancer in humans.

Some farmers in the United States and elsewhere, after nearly half a century of high-technology harvests, are reconsidering older and more natural methods of protecting crops and keeping soil fertile. In nature, plants are mingled together, and help protect one another. In human agriculture, fields are cleared in an effort to raise only one kind of plant, and such fields are thus vulnerable. Farmers are again starting to sow plants in combination (mingling marigolds with vegetables, for example), to discourage certain types of caterpiller moths and other pests. Praying mantises are set loose in fields to devour harmful insects. The use of mulched plants and ripened steer manure is again being used with success on many farms.

A 1974 study headed by Dr. Barry Commoner at Washington University in St. Louis shows that the rising cost of oil now makes such "organic farming" methods cost-effective. Farmers who rotated crops to enrich soil and used natural fertilizers had slightly lower yields and grew fewer profitable crops than farmers using chemical fertilizers, Commoner found, but this was offset by the much lower cost of organic methods.

In some places, like Muskegon County in Michigan, human sewage is systematically treated and returned to the fields as fertilizer, as has been done for millennia in China and India. Why not? "Night soil" is natural and organic, and if not used this way, a good source of food for plants becomes a problem rather than a tool helping us. New York City has for decades piped its semi-treated sewage into the ocean miles from shore, but now the submarine mass of gunk is creeping back toward Manhattan and soon may engulf the Statue of Liberty. Many forms of human waste do more than go away, taking useful resources with them; they also often return to haunt us.

If we could stop crop damage by pests, the world would save 20 to 40 percent of its potential food supply. If we could use natural fertilizers, and practice old-fashioned crop rotation, we could save untold costs for oil-based fertilizer and have the oil available for other uses.

We waste water, which is tragic when global climatic change threatens to bring drought to many countries and irregular rainfall to all. Croplands in the U.S. Midwest depend on rain for water. In China, with less cultivated area, more than twice as much land is irrigated as in the U.S. This does not guarantee China protection from drought, but it improves her chances of surviving years of little rainfall. Irrigation systems are expensive, but they help stabilize food production—and sometimes are required to make it possible at all. Nations confronting hot, dry climates have long been theoretically able to irrigate their lands with desalinated sea water, but desalination is expensive, costing $5 or more per 1000 gallons where fuel must be bought. Even in Kuwait, where fuel is virtually free, desalination costs 60 cents per 1000 gallons; but, as Michigan State University scientist Georg Borgstrom writes, beef-growing requires 50,000 gallons per pound to grow the grass to feed the cow, which means the use of desalinated water would add between $3 and $25 to the cost of every pound of steak. Israeli scientists, however, have helped make desalination economically feasible by developing systems of "drip" irrigation that deliver precise amounts of water to the roots of each plant individually. This also permits the most efficient use of chemicals for the plants, for these can be mixed into the water. The cost of the system is offset in a few years of use by savings in water and fertilizer. Other Israeli researchers have also developed greenhouses which use double panes of glass, between which they inject sea water; sunlight evaporates this water, which drifts inside, condenses on the greenhouse ceilings, and falls as pure water on the plants. These are perfect examples of doing more with less, of human ingenuity as its best.

But we could do more, as Fuller writes. Suppose we viewed humans as "geonauts"—how best could we be equipped to live on this planet? We need what the astronauts use, says Fuller: a life-support packet that would enable each of us to recycle our own waste and water. In a future society we may each use such a 300-pound unit to "supplant the functions now performed by our bathrooms, kitchens, laundries. It will be mass-produced at about one dollar per pound and will be rented for $17 per month." With such units people would be free to live anywhere—on mountain-tops, at sea, in deserts. They would no longer be dependent on the centralized water supplies and sewers of cities for life support.

For the time being, however, history is moving us in the opposite direction from this Fuller vision. The trend now is toward more centralization of facilities; this is in an effort to save resources by economies of scale. The U.S. Atomic Energy Commission developed prototypes of an atom-powered electrical generator, run by safe isotopic decay, which people could keep in their homes to provide all their energy needs without reliance on any outside centralized source. But such generators will never be manufactured or distributed. Instead the U.S. Government is considering creation of ten "regional energy parks" to restrict the pollution of energy production to a few wasteland areas, like Arizona, from which energy will be distributed to the nation. The eastern half of North America is already linked by assorted "power grids" through which regions can share electricity as needed. In November 1965 one small relay in a Canadian power plant within the northeastern grid failed, and for hours 30 million people on the U.S. East Coast were without electricity. In the summer of 1975 a failure in the grid farther north plunged eastern Canada into a similar blackout. Cities, trying to save fuel now lost by automobile commuting, are developing rapid-transit systems. Like power grids, such systems are vulnerable to massive failure, by accident or by sabotage.

Even a city with little rapid transit can be as vulnerable as a moon colony, thanks to dependence on centralized systems that can go awry. Los Angeles, for example, maintains more than 5 million people with water brought by only four aqueducts, by electricity supplied largely by four line systems from as far away as New Mexico. Cut these lifelines off, and social order would soon dissolve. A dozen adept terrorists, working in a coordinated effort, could do this.

At the time of the East Coast power blackout, Pentagon "think tanks" were at work on a project called NES, the National Entity Survival studies. Its goal was to find out how to lower the cost of nuclear war by pinpointing targets in countries that, if destroyed, would kill the nation's sense of identity and purpose, its social cohesion, and set its citizens to fighting one another. The NES computer games quickly spotted how to achieve this in the U.S.: cut off utilities in major cities, and keep them off. The eight Americans in ten who live in cities, it found, were addicted to technology: did their jobs, were tranquilized to sleep at night by six hours of television hypnosis, and slept with a nightlight because they still feared the dark. Moreover, these people tended to be armed, but not to know the people farther away than next door. Urban Americans feared their neighbors, and seemed more at home with television images of famous people in their living rooms.

What happens if the plug is pulled on the artificial reality in which such people live? They quickly develop withdrawal symptoms, fears, and a potential for panic, concluded NES. Panic did not follow when the lights went out briefly on America's East Coast or in Canada, but the Canadian blackout came in summer and ended quickly, and in 1965 the night was warm and lighted by a full moon. Suppose such a blackout lasted days in Los Angeles, where only 2.4 police protect every thousand people. How long would order last before mobs began ruling the streets? NES estimated that such chaos would emerge in between one and two weeks.

Such a collapse would be caused partly by human values and fears and partly by the defective shape of the social structure. Within fifty years terrorists will probably acquire atomic bombs, nerve gas, and other mega-weapons. Modern cities, dependent on centralized technologies, make perfect terrorist targets. Modern city dwellers make perfect targets too. Meanwhile experts talk of unified power grids for Western and Eastern Europe, and even of a global power grid through which nations on opposite sides of the planet can share electricity.

Beginning in 1974, the United Nations has promoted its New Economic Program (NEP), which would allow small nations to take over multinational companies on their soil, would turn all world commodities into cartels, would permit poorer nations to cancel their debts to other countries, and would index the price of raw materials to the price of manufactured goods. NEP would certainly benefit nations with a share of scarce resources, and would spawn a world of "little OPECs," but it would leave nations devoid of resources, like Bangladesh, to starve. Like socialist systems, it would be good at redistributing existing wealth and bad at producing more. NEP would discourage international investment and encourage an isolationist world based on multinational agreements outside the United Nations.

But at this point nuclear weapons proliferation is inevitable. The last attempt to stop it came with the "accidental" deaths of 13 top French nuclear experts only days after their government announced it was going ahead with development of an H-bomb. Now the tide has gotten too high to stop with treaties or selective assassinations of experts or preemptive strikes of the sort the U.S. and Soviet Union in 1963 jointly considered making against nuclear facilities in China.

A world of nuclear powers will demand more equitable distribution of resources and technology. If change from NEP or some other source is not forthcoming, the ugliest sorts of power politics will result as superpower nations seize resources.

The Soviet Union is already preparing for this eventuality. A Soviet army of 2000 coal miners now occupies the potentially oil-rich Norwegian islands of Spitzbergen in the Arctic Ocean, twice the number of Norwegians on the islands. The Soviets could take over Spitzbergen at a moment's notice. Soviet naval wargames have lately concentrated on ways to intercept oil supertanker routes to Japan and North America. Former C.I.A. functionary Miles Copeland writes that U.S. intelligence estimates show a consistent, concerted Soviet policy of trying to cut off U.S. supplies of rare minerals vital to industry.

Of course we may live to see a world in which the centralist logic of NEP will eventually demand the rationing of oil and minerals on the planet, along with land, food, and water.

Curbing Population Growth

And none of these things, good or bad, will mean much in the long run if world population keeps growing. The centralist answer will sooner or later demand mandatory birth control by one means or another. It is almost, but not quite, too late for better remedies.

What else can we do? Again the issue is how to change human values in lands where people now want to have large families. Governments can propagandize. Radio broadcasts can attack ideas of romance and parenthood in a culture, a tactic tried with little success in Egypt. Schools can promote the belief that marriage should be postponed until age 25 or 30, at which time couples tend to have fewer children; where this is customary, as in Ireland, population is stable. Movies and television programs can be produced showing a positive image for celibates, people with small families, or, as *Time* Magazine once recommended, homosexuals. All social and legal reforms making women more nearly equal with men would help, for women would be freer to choose social roles other than motherhood.

Communists can be urged to do as China does, not as she says. China has the strictest population control program of any nation and encourages late marriage, birth control, abortion, and social stigma for large families.

In Roman Catholic nations a new church doctrine might prove useful. At present the Church forbids artificial birth control because it eliminates the possibility of conception with each sex act, and thereby perverts marriage. But the Church apparently does not hold any modern doctrine-of-plentitude view that couples should have as many children as possible, or it would be encouraging use of fertility drugs. On the contrary, the Vatican II Council said parents should desire a family no larger than can be properly cared for given "the material and the spiritual conditions of the times," and Pope Paul VI has said that control of the world birth rate (by "moral means") is desirable.

What if a drug existed that could retard fertility without eliminating it, thus reducing the likelihood that a sex act would produce offspring without removing the possibility. Could Church doctrine accept such a drug? One population advisor to the Vatican tells me it might, but stresses that he cannot speak for the Pope. If approved, and if used only by couples who already had at least one child, such a birth-retardant drug could help countries achieve population control without violation of religious beliefs.

Moveable Feasts

When winter nears, many birds and insects migrate toward the equator. Can humankind do likewise as Earth's climate cools? According to Buckminster Fuller, yes. Cities can exist in geodesic spheres half a mile in diameter that can float in the air, work as greenhouses, be anchored to mountaintops. Larger versions of these "Cloud Nines," a mile in diameter, could be linked so that passengers could walk between them. When winter comes in one hemisphere, the airborne communities would simply move to the other side of the planet.

Similar tetrahedronal cities could float on the oceans, moving with the seasons and supporting tens of thousands of people. They could use atomic power to provide energy and desalinate water. They could grow and harvest sea crops, or catch seafood, or maintain land crops in greenhouses. They could also use deep-ocean pumping methods to provide energy and control their temperatures. And the cost of such mobile cities? Fuller does not venture a guess but claims they are already "real possibilities." Studies by Dr. Richard Meier at the University of Michigan suggest he is right.

Are we prepared to take our caves and fields onto the seas or into the skies? Some scientists say we should move to outer space.

In summer 1975, a group of scientists meeting at the Ames Research Center in California concluded that cities in space are now plausible. They published plans for such a city, a mile-wide wheel-shaped space station able to house 10,000 people; for a mere $100 billion this could be in operation before year 2000. It would be built with materials mined on the moon to avoid the expense of lifting already-scarce Earth resources out of Terran gravity. Its colonists would operate solar energy stations, beaming power by microwave back to Earth, and they would mine precious minerals on the lunar surface. Once established, the space city would be surrounded with 10 million tons of slag from moon mines to protect it against radiation and meteorites.

Princeton physicist Gerard O'Neill proposes an even fancier space station, complete with soil, pastures, rivers, and a system of exterior mirrors that would create for colonists a sense of day and night and seasonal change. Like the island in space proposed by the Ames gathering, he anticipates his space city will orbit between Earth and the moon, in gravitational balance with both.

Such stations in space would provide Earth with energy and minerals that would otherwise have to be obtained by the melting down of old machines. With the advent of the space shuttle, such projects in space will become progressively less expensive. A "seed" world could be put on the moon, complete with a colony of 200 people to mine minerals, by year 1986, at a cost of around $30 billion—roughly what the United States spent on its Apollo program. The minerals dug and alloys created on the moon could soon bring $5 billion in yearly returned goods on this investment.

Someday these stations could serve as stepping-stones for our colonization of Mars and planets outside our solar system.

Many of these ideas smack of science fiction. Certainly carrying them out would involve tremendous energy and effort, and the side-effects and requirements they entail may prove unacceptable.

As environmental climatologist Reid Bryson of the University of Wisconsin commented after having read this chapter, some of these solutions involve terrible problems. "I did a rough estimate of how much polyethelene it would take to put our crops into greenhouses," he said, "and came up with a number like 100 million cubic meters. That's a lot of plastic. And even if we put it up, what would other environmental effects be? Millions of hectares covered with greenhouses would be reflective and would probably change the albedo and climate of many regions." Fusion reactors? They would require vast amounts of minerals to construct, including many times the known world supply of several rare metals. Dome cities? They might become the castles of a new feudalism, producing authoritarian social orders that, through absolute rationing of resources, would punish all dissent with banishment. Most of these solutions, he said, fail to take a holistic view of things. People think of ideas, but not of the physical and environmental costs of carrying them out. There are no cheap, easy, environmentally safe remedies of proven workability now available, Dr. Bryson believes; to find them we must take a broad view and consider the far-ranging implications of new ideas.

Our potential to change, to grow, to adapt to a global climatic cooling is great, but as with our ancestors that potential is tapped only under pressure. Suppose the cooling is the dawn of a new Ice Age. What will that do to our species? "A new ice age need not push us to extinction nor even set civilization back," writes University of Wisconsin geographer Gwen Schultz. "The opposite might be true. It might be the very thing that could revitalize a human race that was growing too pampered, complacent, flabby, and degenerate for its own good. Instead of annihilating us, the Old Cold coming back might annihilate other anti-human creatures—like certain insects or rats or some unsuspected aspirant to world power—and favor us humans as it did several times before. With each of the glacial stages our species became more advanced, more secure, than before."

13

Alternative II

Change Weather and Climate Where We Live

If you catch a cold, the doctor will tell you there is no cure. You must simply wait for it to run its course. But you can use medicines to relieve the symptoms—headache, sneezing, chills and fever, upset stomach, sore throat—and thus prevent its spreading to others while you are awaiting its departure.

In a sense our planet is catching cold, and we have no technology guaranteed to cure this. The climatic cooling will run its course. But we have ways to treat the symptoms of the cold—the storms, drought, frost, shortened growing seasons, heat waves and cold snaps—and thus we can prevent rapid expansion of the cooling and hasten its departure.

Some of these ways have already been mentioned. Cloud seeding can bring rain in drought-stricken lands. It can de-fuse blizzards by making storm clouds drop part of their snow far from population centers—a technique used with some success around eastern Lake Erie and southern Lake Ontario between the U.S. and Canada. And it can reduce hail damage to crops as well, as Soviet research shows. Where needed, such seeding can also be used to create snow, as now is done regularly over America's Rocky Mountains to improve ski conditions and the amount of water supplied by spring runoff.

Climate can be changed when the shape of the land is modified. Various plans have been proposed to blast holes in the mountains north of Los Angeles to help smog and heat blow away, and to dynamite open several mountain passes in Outer Mongolia to allow more moisture-bearing winds to flow into the country. But simpler measures could also be used. During the Dust Bowl drought of the 1930s the United States planted thousands of "shelter-belt" trees in the Midwest to reduce windiness and to slow evaporation. Today the Soviets do likewise on their steppes, and Libya is planting several hundred miles of tree windbreaks to reclaim land from the Sahara Desert. The Rajputana Desert in India and Pakistan "made itself"—where goat grazing and a slight change in climate eliminated native grasses, which left the land bare, which meant reflected sunlight dried the land into a desert unable to re-grow the grasses; today scientists are beginning to cultivate grasses in the desert, having calculated that this will cool air above the land just enough to change the climate again and restore arability to the soil. Livestock overgrazing has also been a factor in the African Sahel drought.

Plants actually have the ability to "seed" clouds, according to a theory by Dr. Russell Schnell of the National Oceanic and Atmospheric Administration. They give off organic molecules that are carried aloft on air currents and which make better condensation nuclei for raindrops than dust. (The U.S. Navy has gone further, suggesting that trees give off molecules that are a prime cause of smog near big cities.) When marginal land is overgrazed and plant life is destroyed, as in the Rajputana Desert and the Sahel, the loss of these organic molecules reduces rainfall and kills yet more green plants. Thus drought kills plants, but an absence of plants can cause drought.

Drought will be a major symptom of continuing climatic instability. In many areas rain will virtually stop coming; in others it will waver irregularly between drought and flood. Cloud seeding cannot always remedy this, especially in regions where increasingly dusty skies already provide an abundance of condensation nuclei. Since 1970, water rationing has been imposed temporarily in Tokyo and Rome and may be needed in Washington, D.C. and other major cities before 1980. Relative to our growing need for it, fresh water is more scarce on our planet than oil, and shortages will occur sooner or later. We use immense amounts of it. For example, each pound of beef, estimates Michigan State University food scientist Georg Borgstrom, requires 50,000 gallons of water to produce, for growing the grass to grow the cow. In industrialized nations demand for fresh water is increasing faster than supply.

The cooling may help solve this problem, if the vision of RAND Corporation researchers is realized. They note that the cooling has already created more icebergs in the North Atlantic. These can be towed near cities and feasibly be used as a source of fresh water; but wherever they sit, the icebergs will reflect sunlight back into space and chill surrounding air. On a hot summer day New York City residents might be refreshed by cool breezes from an iceberg anchored near the Statue of Liberty, but during winter the giant ice cube would be less welcome. Nearer the equator icebergs would be welcome year-round. In 1975 the Saudi Arabian government announced it had hired several French scientists and was launching an expedition to tow icebergs from Antarctica to the Persian Gulf and Red Sea, where they would provide cooler breezes and fresh drinking water. They would also create clouds, which could then be seeded to produce rain. "Over a period," said the head of the project, Prince Muhammad Faisal, "we would hope to change the vegetation and climate in some coastal areas."

For centuries in the Hindu Kush, tribespeople have created frozen reservoirs by breeding glaciers. At the headwaters of tiny streams they bury together two kinds of ice, male (clear) and female (cloudy). After months, perhaps years, the resulting chill turns the stream into an expanding glacier whose water stays put rather than runs off. This wise technology is usable in only a few regions of the world where climate is cool.

New York City would have a more reliable and beneficial water supply if a plan by Lamont Geological Observatory scientist Robert Gerard were used. Gerard wants to dam the east and west ends of Long Island Sound to create a 150-mile-long lake between the northern shore of Long Island and the shore of New York and Connecticut. This water now is salty, but once cut off from the ocean and fed by rain, snow, and rivers, it would become fresh in about seven and a half years, estimates Gerard. New York City would then have one of the world's biggest reservoirs nearby.

A few Canadians have proposed a similar plan—they advise building a dike at James Bay, at the southern tip of Hudson Bay.

This too would soon become a fresh-water lake whose waters could then be pumped into Lake Huron to provide more usable water for Americans and Canadians around the Great Lakes.

Such new lakes would have a climatic impact, but precisely how is uncertain. Bodies of water tend to hold summer heat far into the winter, and winter cold far into the summer; thus they act as climate "flywheels," evening out seasonal extremes of temperature that otherwise would affect lands nearby. But fresh water freezes at higher temperature than salt water, so if the global cooling continues, such reservoirs would turn to ice sooner than they would have when salty, and this could chill winter climate in New York City and around Hudson Bay.

Some artificial lakes have been proposed primarily for their impact on climate, and only secondarily as reservoirs. The Soviet Union has created several inland seas, near the Caspian Sea and elsewhere, in part to provide pools of warmth in winter that will enhance precipitation and moderate climate. Australia has plans to enlarge shallow Lake Eyre, surrounded by desert in the far south central part of the continent. The lake is already salty, so engineers are considering a channel to draw sea water into it. The enlarged lake, they say, would produce enough evaporation to increase needed rain and snowfall over the Australian Alps to the east and over nearby deserts.

Dr. Herman Kahn's Hudson Institute has studied the worth of damming the Amazon River near Obidos, Brazil, to create an inland sea one-third the size of France. The cost of necessary dams would run to about $5 billion, but this would provide vast hydroelectric power and fresh water for the country. Other costs are harder to calculate. The blockage of nutrients the Amazon carries into the Atlantic Ocean would reduce seafood catches as far away as Florida. The flooding of the Amazon Basin would dislocate many native tribes, inundate valuable lands, and doubtless change wind and rainfall patterns. A 1975 study by weather scientists at the University of California's Livermore Laboratory indicates that the average global temperature would drop one-third degree C. if all tropical vegetation on Earth were cut down to open lands for agriculture. Submerging the Amazon rain forests might have similar effects. The United Nations Food and Agriculture Organization estimates that in 1976 at least 400,000 square miles of tropical rain forests will be cut down, an area roughly equal to the proposed Amazon Sea.

Until recently some American engineers hoped to dam the Yukon River northwest of Fairbanks, Alaska. Within twenty years, this Rampart Dam would have created a lake bigger than Lake Erie. Storms that now blow down from the Arctic along the West Coast of North America would have crossed over the giant body of warm water, and presumably this would have improved climate throughout the Pacific Northwest. Construction of the Alaska oil pipeline, which runs next to this region, probably means the dam will never again be considered.

Another unused idea to improve American climate was to create an artificial lake in Arizona that could "make the desert bloom" by cooling climate and enhancing rainfall. The idea was dubbed "Lake Fallacy" by University of Arizona meteorologist James McDonald, who calculated that to increase rainfall by 10 percent over Arizona the lake would need to be 380 miles wide (the width of the state) and 50 miles across (this, too, larger than Lake Erie). Creation of "Lake Fallacy" would require more water than Arizona could expect to get back in rainfall.

One of Egypt's reasons for building the Aswan Dam, which involved the terrible price of cutting off annual Nile floods whose water and silt had kept its northern land fertile for 5,000 years, was the hope of gaining a new climatic region capable of supporting millions of people around the 400-mile-long Lake Nasser the dam made. This lake is long but only a few miles wide, and it has suffered the same fate projected for the hypothetical "Lake Fallacy." Its moisture evaporates and travels over the Sahara on strong winds, but the heat of the desert is such that little or none falls as rain. The region around the Aswan Dam remains arid.

Awesome as the Sahara Desert seems, plans have been put forth to alter its climate. In 1912 a French professor named Etchegoyen proposed digging a 50-mile-long canal from the Atlantic Ocean to flood parts of the desert below sea level with 35 trillion tons of water and produce a "Sea of Sahara half the size of the Mediterranean." This, he told a reporter from *Scientific American*, would make the Sahara "as fertile as Europe." Experts warned that the idea, while feasible, might improve Africa while burying Europe under a layer of permanent ice (France occasionally is covered with veils of windblown Saharan dust, and likewise would be moistened), or might tip the axis of the Earth, or might alter world sea levels.

Better known is the "Atlantropa Plan" first conceived in the 1920s by German civil engineer Heinrich Soergel. Today the Medeterranean Sea is a water sink. Thanks to hot winds from the south, it annually loses more water to evaporation than all its rivers, including the Nile and Po, can replace. The deficit is made up by waters flowing into the Mediterranean from the Atlantic and the Black Seas.

Soergel's plan was to seal off the Mediterranean Sea by damming the Dardanelles and the Strait of Gibraltar. This, he said, would reduce the water level in the Mediterranean by four and a half feet per year. One-hundred-forty years after the dams were built, it would be 660 feet lower than it is today, and 150 million acres of fertile land would be available where the sea once was. The remaining Mediterranean Lake could then be stabilized at a desired level if some Atlantic waters were allowed to spurt in at a controlled rate through an electricity-producing turbine. A variation of the plan called for the building of another dam across the center of the Mediterranean; electricity would be generated if the halves were kept at different levels and the flow between them tapped.

A 1959 analysis by American oceanographer Henry Stommel suggests that the Gibraltar Dam might be a bad idea. The Atlantic Ocean is far saltier than the Pacific, in large part because of the salty flow into it from the Mediterranean. The saltiness of water determines how dense it will be at different temperatures. At present, Arctic waters in the North Atlantic sink to the bottom because of this density phenomenon, but if the Atlantic were no saltier than the Pacific Ocean these icy waters would flow near the top of the North Atlantic. If this happened, said Stommel, the eastbound waters of the Gulf Stream would never reach Norway or the Arctic. Northern Europe would become much colder.

The Atlantropa Plan, in addition to damming the Gibraltar Strait, would bring surface water from the Congo River region of Africa by canal or aqueduct into the central part of the Sahara Desert. This would create a lake four times the size of France to parallel the Mediterranean Lake. Between the two bodies of water, and around them, climate would become humid. Rain would be plentiful. Climatologist Hermann Flöhn, former director of research for the German Weather Service, says that such a lake in the Sahara might have no more impact on climate than does the Aswan Dam. Evaporation, he notes, does not guarantee rainfall.

In recent decades, the Sahara Desert has been marching south at up to 30 miles per year. The most likely reason has been the global cooling trend. Grass and bushland in the Sahel region, along the desert's southern edge, have retarded the expanding sand for centuries. But the cooling of Earth's climate has pushed the atmosphere's monsoon rain belt toward the equator. Deprived of this rain, and overgrazed by nomadic herdspeople, the grass and brush of the Sahel shrivelled. In the drought-stricken area the desert has grown quickly. An estimated 200,000 people in the Sahel died of thirst and famine between 1971 and 1974.

Late in summer 1974 some rain fell in the Sahel, and the world rejoiced that "the drought was broken." But within the American defense community there were rumors that the Sahel rains were not a natural change, but the product of secret weather modification. Those specialists I questioned were quick to admit their suspicions were based mostly on circumstantial evidence, but the case they presented was intriguing: The drought and famine in the Sahel was an embarrassment to the world superpowers. Harvests in the Soviet Union had been bad in 1972, and in 1974 the U.S. crop was shaping up as a disaster. Leaders had been reluctant to pour food aid into the Sahel, fearful for their own reserves, and only when the famine was far advanced did they act.

But the Sahel was a political and economic threat to the superpowers too. It had become a symbol for those scientists who warned that Earth's climate was cooling, tangible evidence of what might soon befall all nations of the monsoon rain belt. If traders on the world commodity markets or fanatical politicians in the small nations started to recognize this danger in the Sahel drought, all hell could break loose in the ensuing panic of small countries to hoard food grains or gain atomic weapons. As a symbol the Sahel was a problem, agreed the U.S. and Soviet Union, and something had to be done—something more important than merely feeding the hungry.

The Sahel is perhaps the most "marginal" of Earth's heavily populated lands. Its climate is so delicately balanced between drought and survival, says Massachusetts Institute of Technology meteorologist Jule Charney, that the advance of the Sahara over it was probably a biofeedback phenomenon; the desert came where failure of the monsoons killed plants, and stayed only because it prevented plants from regrowing. U.S. analysts concluded that if rain could be brought back to the Sahel by some trick of technology for only a year or two, the grass would grow again and the desert would be permanently pushed back—assuming the monsoons returned soon. Even a short respite from the drought would buy political time, because it would discredit those scientists who warned that the Earth was cooling and the monsoon regions were in danger.

As part of the ongoing Global Atmospheric Research Program (GARP), a tropical experiment was planned for 1974 involving ships and planes from several nations. It was to take place in the mid-Pacific Ocean. But under urging from U.S. and Soviet scientists and government officials, its location was switched, at the last moment, to the mid-Atlantic and called the Atlantic Tropical Experiment (GATE).

Late in July 1974, forty research ships from ten nations (including fourteen Soviet ships) began to assemble in the Atlantic for GATE. Less than a month later the first significant rain in eight years fell in the Sahel.

In 1959 Swedish meteorologist Tor Bergeron had theorized how rain could be produced in Africa at 10 to 17 degrees North Latitude, the climatic belt of the Sahel and the Sudan. If the Congo River could be rerouted across slightly higher land and fed into Lake Chad, and if additional regions around the south Nile and the Niger rivers were extensively irrigated, the additional rainfall resulting from evaporation would turn this marginal climatic belt into a garden able to support many millions of people. If irrigation was not used thus, Bergeron said, rain in the Sudan and Sahel could also be produced by evaporating sea water near the coasts of the Gulf of Guinea, boiling the surface water there with the heat from atomic reactors if necessary. A decade later Dr. Flöhn wrote: ". . . the meteorological-climatological basis of Bergeron's proposal seems to be sound," but "the political problems would, perhaps, be more difficult to solve and would require effective and very close co-operation between the emergent African states."

Some of the outriding ships of the GATE experiment were in

the Gulf of Guinea. Did they use one of the several tools developed for speeding evaporation of sea water, such as giant plastic chimneys, to spurt fountains of water and steam skyward? Did they coat the sea with special detergents developed by the U.S. Defense Department to hasten evaporation and form clouds? If such things were done, weather modification has reached a new level—for this means the clouds were grown and directed and harvested, not merely hunted down; the advance would be comparable to the change in human history from hunting to agriculture, the new development meriting a new name like aeroculture or cloud farming.

Two things seem dubious about this line of reasoning. One is technical: rains returned to the Sahel in August, but as Flöhn notes in analyzing the Bergeron proposal, typically during July and August the net air transport of water vapor in the African climate belt 11 to 12 degrees North Latitude is shifted west toward the Atlantic Ocean. (The heart of the Sahelian drought was in the latitude belt 14 to 18 degrees North.) Thus it is uncertain whether increased evaporation in the Gulf of Guinea could have produced much rain in the Sahel, unless the winds of the region also were changed by the global cooling. The other reason is practical and political: no one involved—the Soviets, the Americans, the British, or the Canadians—could have done very much during GATE without the others knowing it, so we must assume either that nothing was done or that whatever weather modification was attempted had the mutual consent and cooperation of all these nations and of many independent-minded, ethical scientists, which seems unlikely, even absurd, unless one fancies conspiracy theories of history.

Given a minimum of cooperation among the participating superpowers, however, GATE provided a perfect cover for clandestine weather modification. Research aircraft of several nations were darting throughout the tropical Atlantic region and above the nations of Africa's East Coast, unrestricted because they were chasing clouds and air currents in the name of international peace and cooperation. Such aircraft could easily have carried out widespread cloud-seeding operations with little fear of detection. Months earlier the U.S. government had refused a request for cloud seeding from Sahelian nations, perhaps out of fear that if the seeding failed the U.S. would be blamed and if it succeeded Americans would be accused of stealing rain from adjacent countries. But hidden under the shadow of GATE, such cloud seeding could have been done without fear of controversy. Both the U.S. and the Soviet Union had motives for trying to produce rainfall in the Sahel and had opportunities. This does not prove it was done, but it suggests that the possibility cannot be wholly ruled out until a study of satellite photographs made during GATE is done to determine where the Sahelian rains came from. The rains that fell that summer alleviated the suffering of several million people, at least for a time.

In 1951 the Nobel-laureate meteorologist Dr. Irving Langmuir proposed that the U.S. should begin regular, massive cloud seeding of the world's trade wind belts, particularly near Hawaii and Puerto Rico. This would cause increased rainfall and "profound changes" in global winds of benefit to all nations farther north, he said. Certainly it could alter weather farther north, making dry areas wet and wet areas dry.

The government ignored Langmuir's advice, but the Pentagon has considered many similar ideas. Studies have been done on the viability of seeding specific wind patterns to ruin an enemy's weather, and of spreading chemical dust in the atmosphere over an enemy nation to blot out sunlight or intensify the greenhouse effect and cause droughts. Tests have indicated that the jet stream can be deflected by putting the right sorts of dust into the atmosphere near its path, or by using chemical coatings on ocean surfaces beneath it to speed or slow evaporation, or by heating or cooling spots of ocean surface by other means, such as oil fires. Such thermal forcing acts like a new set of invisible mountains on the planet's surface and can change wind currents. As the jet stream moves, so does the circumpolar vortex and with it the climate. These methods could be used to alter the jet stream to the advantage of one nation or the disadvantage of another.

Such methods can be costly. A film of hexadecanol big enough to coat 1000 square miles of ocean, but with a film only .001 mm thick, might cost $2000 million. Wind, ocean currents, and storms would make it hard to keep intact. These, plus initial problems of dispersal, also limit a nation's ability to spread massive dust clouds in the atmosphere, or to cover the Arctic icepack or the world's deserts with soot or coal dust to reduce the amount of sunlight they reflect.

But massive seeding of wind patterns with silver iodide generators, and to a lesser degree the thermal alteration of spots of ocean surface, are quite possible. They are both financially and technologically viable policies for a wealthy nation.

Perhaps the part of the world with the least rainfall and the most evaporation is the Red Sea. In 1960 an Egyptian oceanographer, Dr. El Sayed Mohammed Hassan, proposed damming the south end of that sea and putting a small barrier with sluices at the Suez Canal. This would prevent a direct flow of water into it from either the Indian Ocean or the Mediterranean Sea. Hassan estimated that 120,000 cubic feet of water evaporate from the Red Sea every second, and that it would quickly dry up if it were dammed. Dr. Hassan's plan would not let the Red Sea disappear, but would use it for electrical power. Indian Ocean water would still flow into it through turbines in the dam. This, he estimated, would generate power equivalent to one-fifth of the total 1960 world output.

The Red Sea Dam would provide cheap electricity for all nations in the region, even Israel, said Hassan, who believed his dam would create a billion jobs in Africa and the Middle East. With it desert nations as distant as Libya could power ocean desalination facilities, pump water into irrigation systems, and make the desert bloom. Thus, abundant energy could alter topography of the region, and with it climate. (In the dam's absence Libya is operating atomic desalination facilities and benefiting from the global cooling, which has already pushed the Sahara far enough south to moderate Libyan temperatures and increase rainfall.)

Heat is a problem in some places, but prayed for in others. On the east coast of Siberia the Soviets have long contemplated building a dam across the Tatar Straits, which stretch less than 100 miles at the widest point between Sakhalin Island and the mainland. South of the straits is the warm Sea of Japan; to the north, the cold and foggy Sea of Okhotsk. Every day, when the tide rises, warm water surges north into the straits, and daily it surges out again as the tide falls. If only it could be made to keep moving north, Soviet engineers decided, Siberia could be warmer.

The dam they devised was simple: a series of six huge gates, each 330 feet wide, would open as the tide rose and close as it started to ebb. These one-way gates would have the effect of creating a slow manmade current of water moving north through the straits. Nearly 40,000 cubic feet of this warm water would flow north from the dam each day, estimated Soviet planners, and this would gradually warm coastal temperatures in the Sea of Okhotsk by 10° Celsius.

Japan issued a strong diplomatic protest when the Soviet idea became public. If the Sea of Okhotsk warmed, they said, the course of the warm Kuroshio Current, the Pacific Ocean's "Gulf Stream," would change and Japan's climate would turn colder. The Soviet plans for the Tatar Straits dam have been shelved, at least for now. Concern that similar political problems might arise has apparently discouraged European engineers who during the 1950s talked of damming the English Channel with gate dams that would produce warming currents between Great Britain and the continent.

Many methods can be used to ameliorate effects of the global climatic changes in parts of the world, but all will involve political difficulties. The weather- or climate-modification technologies used by one nation will almost certainly change weather in neighboring nations, or will be blamed for nearby weather changes caused by the unstable climate. Political protests in 1912 and 1913 discouraged the U.S. Congress from funding research into Carroll Livingston Riker's idea to dam the Labrador Current between Newfoundland and the Grand Banks. Fear of such protests has also shelved a Canadian plan to put an atomic reactor near Hudson Bay to heat its waters, because though it might warm the region around the Hudson Bay, it could also increase snowfall around the Great Lakes and elsewhere.

As early as 1950 U.S. Department of Defense think-tanks were considering whether to use nuclear explosives to blast a new sea-level canal across Central America. Their conclusion: Such a canal would create a flow from the higher, cooler Pacific Ocean into the Caribbean Sea, and this could weaken the Gulf Stream flow—thereby chilling the U.S. East Coast and Europe. (Their peripheral conclusion: the most effective way to freeze Europe would be a simple 50-mile-long dam across the Straits of Bimini, birthplace of the Gulf Stream.)

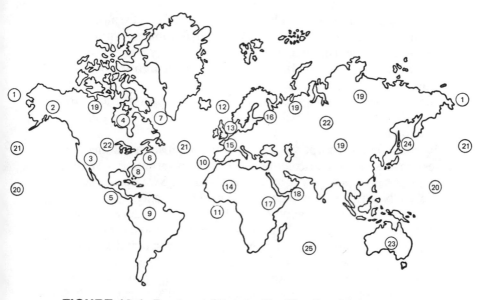

FIGURE 13-1 *Regional Climate-Modification Ideas*

1. Dam Bering Strait
2. Dam Yukon River to create giant lake
3. Create "Lake Fallacy" in Arizona
4. Artificially heat Hudson Bay or dam its southern tip at James Bay
5. Blast sea level canal across Central America
6. Dam Long Island Sound
7. Dam Labrador Current
8. Dam Gulf Stream at Bimini Strait
9. Dam Amazon River to create inland sea
10. Dam Strait of Gibraltar
11. Use heating or other means to increase evaporation in or near Gulf of Guinea
12. Dam or blast submarine ridges near Norwegian Straits
13. Dam English Channel
14. Divert ocean or river waters to create inland sea in Sahara
15. Dig canal from North Sea to Mediterranean Sea
16. Dig canal from Arctic Ocean to Baltic Sea
17. Dig canal to eliminate swamps from upper Nile River in Sudan
18. Dam Red Sea
19. Divert Arctic-flowing rivers southward or dam them
20. Cloud-seed the trade winds
21. Heat or cool spots of ocean surfaces to alter global winds
22. Put stationary dust clouds over enemy nation, either in atmosphere or in outer space
23. Dig canal to ocean to turn Lake Eyre into inland sea
24. Dam Tatar Strait between Sakhalin Island and Siberia
25. Tow icebergs from Antarctica.

Other canals have received less scrutiny. The Soviet Union, for example, in 1975 completed a new, deep canal from the White Sea in the Arctic to the Baltic Sea. Big enough to carry destroyers, it probably injects brackish Baltic water and heat into the Arctic Ocean. What climatic consequences will it have? Likewise, what of the French and German canal system now being constructed to link the North Sea to the Mediterranean? Its environmental impact is unknown.

Egypt, Kuwait, and perhaps several other Moslem nations are investing heavily in construction of a canal to reroute the White Nile in Sudan. Their goals: to dry up the Sudd Swamps through which the Nile flows; to make the Nile navigable; and to provide a system of controlled irrigation that will make the Sudan a "bread basket" for the Moslem nations of the region and end their dependence on food imports. But environmentalists warn that drying up the Sudd Swamp will radically alter rainfall over at least a third of the Sudan region and perhaps farther away. Syria, meanwhile, is damming the other major river of the cradle of civilization, the Euphrates, and this will modify climate in the whole of what once was Babylon.

Using regional climate modification to treat symptoms of the worldwide climatic change involves unknown environmental risks and known political problems, but as weather worsens the political climate will change and the risks will seem increasingly acceptable. The effects of climate-modification technology inevitably will extend somewhat beyond the lands it is used to change. Because these effects will tend to follow the paths of wind and ocean currents, we can assume that weather and climate modification will encourage a new kind of political alliance. Nations in the latitude zone 10 to 18 degrees North, which depend on monsoon-type rains, will discover they have a common political interest in efforts to gain rainfall. Canada, the Soviet Union, the United States, and Northern Europe—as regions that will suffer most on the planet if the cooling continues unchecked—may find cooperation in studying and modifying regional climate in their combined self-interest. Wherever nations share wind and ocean currents, so too do they share concern for what changes occur in those currents, natural and manmade.

14

Alternative III

Try to Control Global Climate

The world can be warm for thousands of years, but then the ice near our poles begins advancing. The world can be buried in an Ice Age for thousands of years, but then suddenly start warming. Why? Perhaps our sun changes the amount of energy it puts out, turning hotter or cooler over long periods of time. If this is why the ice comes and goes, writes Dr. Gwen Schultz of the University of Wisconsin, "there is nothing we could do to forestall it."

But what if the Ice Ages are caused by some change on the Earth's surface? If this is so, we might be able to discover what it is and, by preventing it, take control of our planet's climate as surely as we now control temperature in our homes with a thermostat.

In 1956 two scientists at Columbia University's Lamont Geological Observatory said they had discovered why Ice Ages happen, and that the key was the Arctic Ocean. Two decades earlier British meteorologist Sir George Simpson theorized that Ice Ages occur when the sun burns hotter; this increases the amount of moisture-laden air arriving at the North and South Polar regions, where cold conditions convert the moisture into snow and ice. Sir George was partly correct, Maurice Ewing and William L. Donn seemed to say. The sun may not have burned brighter, but more moisture in the polar air *was* the reason Ice Ages occurred. This moisture, however, came not from the tropics as in Simpson's theory, but from the Arctic Ocean during periods when it had little ice.

215

Before the present Ice Epoch, which keeps part of our planet covered with ice continuously, Earth was warm and ice-free. In those days, theorized Ewing and Donn, the North and South Poles were not where they now are. A few hundred million years ago Earth had only one giant continent of land, but movement in the planet's crust broke this huge landmass, called "Pangea," apart. The pieces, each of them a giant hunk of rock floating on molten rock in the Earth's core, began moving apart. This "continental drift" today has moved the pieces, which we call our continents, to their present far-apart positions, and continues to move them. The pressure this movement puts on the tectonic plates of Earth's crust is periodically released in earthquakes.

When Pangea was first breaking up, its pieces separated at what we now call the tips of South America, Africa, India, Australia, and Antartica, which to this day could fit together almost as nearly as a puzzle. Even before the ocean rushed in between the parting pieces, this region was gripped by the long-lived Permian Ice Epoch 250 million years ago. Why? Perhaps because Earth's South Pole was then near this separation point. The region was cold from lack of sunlight during much of the year, but warm enough that water vapor from the new ocean waters could evaporate and feed snowfall.

The continents drifted away from the South Pole after the Permian Ice Epoch and the world warmed. Dinosaurs emerged and ruled the world while 120 million years of warmth lasted. The world was ice free then, suggested Ewing and Donn, because open ocean existed over the poles. Thus, although the poles were the coldest points on the planet, ocean currents easily carried heat to them from the tropics. If any snow fell, it landed in the warm waters and melted quickly.

Then the continents moved again, and this set the stage for the Pleistocene Ice Epoch from which humankind emerged. The continents shifted around the Arctic Ocean, cutting off this small body of water from tropical ocean currents. The continent of Antarctica settled over the South Pole, providing a landmass where snow could settle and turn to ice. Thus both poles grew colder. Global climate cooled. Soon the first ice age of the new Epoch sent ice sheets expanding across the lands bordering the Arctic Ocean—the Soviet Union, Greenland, North America, and Europe. For a time it must have seemed the Earth would be locked in ice forever.

But then, wrote Ewing and Donn, the Ice Age destroyed itself. When the world got cold enough the Arctic Ocean froze over. This cut off the evaporation that had fed the snow's advances. Unable to grow or renew themselves, the giant fields of ice halted and began to shrink. At last the ice retreated to near the Arctic Circle, where it achieved a seasonal equilibrium. Every summer some ice would melt from the Arctic Ocean. Every winter, as the sun hid on the other side of the planet, the frigid Arctic night would squeeze any evaporated water back into snow and ice, and the ocean would refreeze.

Over the long term, however, the equilibrium was imperfect. Hundreds of subtle natural feedback mechanisms changed with the climate, and they imparted to such changes a pendulumlike momentum. Once a cooling or warming trend began, it continued until some large force stopped it. Once the ice began retreating poleward, it set in motion other forces that would cause a warming trend.

Foremost among these forces was sea level, said the two scientists. If Ice Ages grow by converting ocean water into landlocked ice, then as long as the ice expands, sea levels will drop. During the last Great Ice Age world sea levels fell 250 feet lower than today's. When they fell they guaranteed the end of the Ice Age.

Why? The answer is amazingly simple. The Arctic Ocean is not entirely surrounded by land. It connects to the Pacific Ocean through the Bering Strait between Alaska and Siberia. It connects to the North Atlantic through the Denmark Strait between Greenland and Iceland, and through open ocean between Iceland and Norway. By all these connections the Arctic Ocean receives warm currents swept north from the tropics—especially via the Denmark Strait, through which Gulf Stream waters reach the Arctic.

But all these connections are shallow and flow across undersea plateaus. The average depth of the Bering Strait, for example, is 150 feet. The other links between the Arctic and warmer oceans are little deeper. Thus when an Ice Age gathers momentum, and its snowstorms cause global sea levels to fall, the warm-water flow into the Arctic is soon cut off. During the last Ice Age the Bering Strait rose above sea level, and for thousands of years it served as a giant land bridge across which the ancestors of American human beings, animals, and even plants migrated eastward. Likewise, the Denmark Straits and other regions in the North Atlantic rose near or above sea level; ice grew over them, and whatever currents from the south prevented the Arctic Ocean from freezing were blocked. For a time the Ice Age continued on its own frosty momentum, but with the closing of the Arctic Ocean and its consequent freezing, the fate of the Ice Age was sealed. Its prime source of fuel was gone, and it was doomed to retreat.

As the ice melted, sea levels again rose, submerging the Bering land bridge and the Denmark Strait. Warm ocean currents resumed their flow toward the Arctic, and this speeded world warming. The ice tried to hang on, clinging to cold mountains year-round and enduring as an unyielding mantle over Greenland, Alaska, Siberia, and the Canadian Arctic. Its retreat slowed but did not stop. And with every bit of ice that surrendered, a bit more water flowed back into the oceans and raised their levels. With every inch the oceans rose, the channels between the Arctic Ocean and warmer oceans deepened and could carry more warm water. Eventually, wrote Ewing and Donn (whose widely debated theory has been grossly simplified in the above description), enough warm water reached the Arctic to melt most of its ice pack. The lands around the ocean were still covered with ice, and when evaporation started to rise in abundance from the newly defrosted Arctic Ocean it fueled new snowfall and a new advance of the ice began. In a cycle that will never end until Earth's poles again shift location, the ice grew, the sea level dropped, the cut-off Arctic Ocean refroze, and the ice retreated until, once more, sea levels rose and the Arctic Ocean thawed.

The theory seems able to explain why Ice Ages might happen, and why they occur in regular, even cycles. Many scientists have noted flaws in evidence, and Ewing and Donn have revised their theory many times to refute critical objections. Did the continents really shift location, and from where to where? Evidence puts the Ewing-Donn theory in doubt on this point. They reply that data could be in error by as much as 1,200 miles in pinpointing where the poles have been. Evidence also indicates that the Arctic Ocean has not been free of ice for one million years, but that in that time several Ice Ages have come and gone. Ewing and Donn originally wrote that glacial periods began when the Arctic was "ice-free," but by 1966 they revised this and now say that much of the Arctic was free from ice, but not all. They predicted, as did Simpson, that oceans should be warmer during Ice Ages. A study of sedimentary layers from the bottom of the Norwegian Sea, one of the connections between the Arctic and the Atlantic Oceans, was released in 1975; it provides evidence that during phases of the last Great Ice Age, as assumed by Ewing and Donn, the sea water was 1°C. warmer than it is today. Soviet climatologist Mikhail Budyko has reported a retreat of ice in the Arctic Ocean since the recent warming trend in world climate neared its peak in the 1930s. According to the Ewing-Donn theory this should cause global cooling through increased snowfall. In August 1975 British climatologist Hubert Lamb and three colleagues reported trace warming in the ocean surfaces of the North Atlantic from 1971 through 1974. Perhaps as a result, snow cover in the Northern Hemisphere increased 12 percent during those years, and winters grew nearly a month longer. World temperatures have been falling rapidly since the 1940s. Despite their flaws, the theories proposed by Sir George Simpson and by Drs. Ewing and Donn are able to explain the seeming contradiction between thawing in the Arctic Ocean and the global cooling climate better than most other schemes can.

The Ewing-Donn theory is wrong, most scientists today agree, because it has one fundamental flaw. In 1957 University of Miami scientists Cesare Emiliani and Johannes Geiss were first to notice that the chain of events it describes requires "capacitance and impedence" in the Arctic so warmth can be retained, but these are lacking. As Emiliani and Geiss explained, when snow starts building up on land at the dawn of an Ice Age, sea levels begin dropping *immediately*. Thus the volume of warm water reaching the Arctic starts diminishing *immediately*. Assuming that as soon as its ice thawed the Arctic Ocean began feeding snowfall on land, a tiny drop in temperature should cause refreezing, and Arctic temperatures should begin falling *immediately*, at least during the months of Arctic night that come each year when no warming sunlight arrives. Thus the Arctic should freeze at the outset of each Ice Age and remain frozen while it lasts. Evidence shows the Arctic Ocean was frozen during the last Ice Age.

But some natural mechanisms set the ice advancing and retreating. Because our power is small, our only hope of stopping an Ice Age lies in altering some tiny but essential part of Earth's climate machinery. With the exception of Milankovitch's analysis of cyclic changes in sunlight, over which we have no control, no single present theory adequately explains why the ice comes and goes. But a theory is needed before we can judge the merits and risks of any plan to modify global climate.

As an example of why this is so, we will consider one of the best-known ideas for controlling Earth's climate as it would be evaluated using the discredited but widely-publicized Ewing-Donn theory. According to Ewing and Donn we can control world climate by regulating the melting of ice in the Arctic Ocean. The more it melts, the more danger we are in. What can be done to prevent such melting?

The Bering Dam

In their long quest for good ocean harbors and relief from a severe, cold climate, the Russian people have for centuries looked at the Arctic Ocean with despair. Following the Russian Revolution the new Soviet leader Vladimir Lenin looked at it with hope. One of his first acts as head of the new "scientific Marxist-socialist state" was to dispatch research teams to study water and weather conditions in the Kara and Barents Seas in the Arctic. Early reports encouraged Lenin's belief that science could be used to master the Arctic.

In 1921 Lenin announced that he wanted to build a dam-bridge across the Bering Strait between Siberia and Alaska, and appointed academician Gleb Krshishanovsky to make plans for the structure. A major benefit of the dam-bridge, said Lenin, would be the high-speed electric railroad that could run across it. This, said advocates of the idea, would let a person travel from Paris to Miami, Florida via Moscow and Anchorage. But the Soviet Union was torn by internal political dissent and economic problems. When Lenin died, for a time it seemed the Bering Dam idea had been interred with him. During Joseph Stalin's reign ideas for modifying climate were still discussed among scientists and engineers, but the Bering Dam project had little popularity compared to schemes for giant inland dams and methods for reversing rivers that flowed into the Arctic such as the Pechora, Yenisey, and Ob. Results from a Bering Dam were uncertain, and such a dam would require large amounts of money and labor and cooperation from the owners of Alaska, the United States. It seemed a bad idea.

But in 1956, shortly after Stalin's death, the Bering Dam scheme was resurrected by Soviet planners. In the magazine *Novy Mir*, a leading Soviet civil engineer, Arkady Markin, went a long step beyond Lenin's proposal. Markin called for a dam equipped with giant atomic-powered pumps. These pumps would suck warm water up from the Pacific Ocean's equivalent of the Gulf Stream, the Kuroshio Current that flows northward past Japan, and push that warm water into the Arctic Ocean. This would melt ice along the Soviet's northern coast, said Markin, and improve climate where Kuroshio waters intensified along the east coast of Siberia. The Soviets had never lost their wish for a land link to North America; in 1946 Soviet academician Vladimir Obraztov, with Stalin's approval, called for a tunnel under the Bering Straits that could carry "chains of motor cars streaming back and forth." Markin reasserted Lenin's vision of a railroad, and perhaps a highway as well, across a Bering Dam.

Markin proposed a dam 55 miles long, and with a cross-section of more than 36 million square feet—by far the biggest dam ever considered as of 1956. But as Markin said, his plan was not just for a dam but for a global "climate factory" that would melt the frozen lands of the world and make them usable for agriculture and commerce. Moreover, the project offered the U.S. and U.S.S.R. a valuable joint venture that would promote cooperation and good will between them, a chance to thaw the political Cold War as it melted ice in their frozen northern lands. And although gigantic, such a dam was buildable, U.S. engineers agreed. Water in the Bering Strait was only 150 feet deep.

In North America Markin's proposal was ignored by most commentators and experts, and attacked by a few. The Toronto *Financial Post* expressed concern that such a dam would increase the amount of icy Arctic water in the Labrador Current that flows down the Maritimes' coast of eastern Canada. This, the paper guessed, would shorten the growing season in that region, create more storms and early frosts, and perhaps freeze up the port of Halifax during winter.

But, wrote the *Post*, a simple remedy existed. Atomic heating stations could be set up in the Arctic to offset this cold-current flow. Tongue in cheek, *Fortune* magazine said editorially that such heating stations would hasten melting of the already-deteriorating Greenland ice pack. If Greenland thawed, said *Fortune*, "the present level of the world's oceans, some scientists estimate, could rise five feet—enough to flood substantial portions of New York, London, the Netherlands, and hundreds of Atlantic seacoast cities and towns. Though the *Post* halts its interesting article at this point, the answer to this objection is pretty obvious, too. What's to prevent the atomic air conditioning of Greenland?"

"Of course, the idea of creating a 'climate factory' in the Bering Strait needs further examination and careful scientific calculations," said Markin. "All this could be considerably speeded up by pooling the joint research work of engineers and scientists of the countries interested." The United States was not interested at the time in such an undertaking.

In 1958 another Soviet engineer, Pytor Mikhail Borisov, proposed a new version of the Bering Dam idea to the Polytechnicum in Moscow. Like Markin's dam, it would feature atomic-powered pumps, but in Borisov's dam the pumps would take cold water from the Arctic and push it into the Bering Sea, and hence the Pacific Ocean, at the rate of at least 500 cubic kilometers each day.

Borisov's reasoning as to why water should be pumped out of, rather than into, the Arctic was remarkably similar to American engineer Carroll Livingston Riker's analysis of ocean currents more than four decades earlier. By pumping water into the Pacific, said Borisov, the Labrador Current, which flows south into the Atlantic, would be considerably weakened. The Labrador Current is nature's prime force in deflecting and weakening the eastward motion of the warm Gulf Stream. The two currents collide in the North Atlantic. If the Gulf Stream retained its power, said Borisov, it would flow across the Atlantic and into the Arctic Ocean. The Bering Dam, by sucking water from the Arctic, would encourage this flow even more. With luck the Gulf Stream could be drawn continuously across the northern coast of the Soviet Union, warming sea, air, land, and weather as it went past. The Soviets would have created their own favorable, manmade ocean current, and with it their own beneficial climate.

The new influx of warm water would quickly melt all ice in the

Arctic, said Borisov. Without the ice, the open ocean would soon be absorbing from sunlight more than "150 times the heat of all coal, oil, and gas now annually produced throughout the world." This, he told the Polytechnicum, would raise the average temperature at the North Pole "by about 27° Fahrenheit. This rise of temperature would lengthen the warm season. Important polar regions could be given over to agriculture, and the rivers of Alaska, Siberia, and Canada would be navigable the whole year round. New agricultural areas will have been created."

Perpetual frost covers 47 percent of the Soviet Union, and 70 percent of Alaska, Borisov told readers of *Literaturnaya Gazeta* in 1959. If the United States and Soviet Union joined hands in creating a Bering Strait dam, this permafrost could be melted, and both nations would benefit. These northern lands would become a "garden," he said. "The whole shore of the Arctic Ocean in January would be approximately the same temperature as the Ukraine and ships would be able to navigate throughout the entire year."

Moreover, wrote Borisov, many "scientific workers and institutes in the Soviet Union, on studying the plans for the Bering Strait dam, think its construction is quite possible." Borisov envisioned a dam 46 miles long, from Chutkotka, Siberia, to the Prince of Wales Cape in Alaska. The dam would be anchored on several thousand concrete caissons that would be built on land, towed into place, weighted down with mud, and sunk. Concrete walls would then be mounted between the caissons. His estimated cost for the dam: $17.5 billion (1959 dollars), or "just about twice as much as the Soviet Union has spent in past years on making her far eastern provinces fertile."

Critics strongly questioned Borisov's ideas about the dam. Soviet meteorologist Dimitri Drogaytsev warned, "If we were to warm the Arctic it could be that the climate would deteriorate in the south instead of improving. The summers would get even drier and the winters colder. True, we would get luscious meadows in the far north, but aridity would presumably also creep further north." At worst, he said, the dam might do "irreparable harm" to Soviet climate by expanding the Gobi Desert north into fertile areas, and by creating ice floes that would clog Siberian rivers emptying into the Arctic and pile up around the dam itself. Such ice against the northern coast, he said, would cool the climate, not warm it; winters would be colder, harbors icier.

In the U.S defense community, Borisov's plan was viewed with horror. Since the early 1950s military intelligence had been aware of Soviet schemes to pump icy Arctic water down along America's West Coast. The goal of such schemes was to cool the climate in North America, increase storminess, and hurt agricultural production, all of which would damage the Canadian and American economies. Borisov's plan called for pumping Arctic waters into the currents that flow down this West Coast, so it was instantly suspect and loudly condemned by Pentagon policymakers in the highest offices of the U.S. government.

Dr. Harry Wexler, the U.S. Weather Bureau's chief of meteorological research, said more scientific understanding of weather and climate was needed before a wise decision about such a dam could be made. Gordon Lill, head of the geophysics branch of the Office of Naval Research, said he doubted such a dam could warm the Arctic. R. A. McCormick of the U.S. Weather Bureau said such a dam could certainly melt the Arctic icepack, but only after 100 years of pumping. Dr. Wexler warned that improved climate in the Arctic might make Southern California and the Mediterranean, which now depend on Arctic-spawned storms for their winter rainfall, "hopelessly arid." More recent analysis suggests it would disrupt weather patterns in Brazil, Zaire, and a host of other nations as well.

But Borisov's proposal won support too. Alaska had just become a state, and its new Senator, Ernest Gruening, believed the idea "should be studied seriously." He prodded fellow lawmakers into discussing the plan, although none supported it openly as he did. "The Soviet Union," said Gruening, "is offering us the possibility not only of melting the ice of the North Pole, but the ice of the cold war as well. Carrying it out would open up a new continent and be a blessing for all the lands of the Northern Hemisphere."

During the 1960 American presidential campaign, the journal *Bulletin of Atomic Scientists* submitted questions to the two leading candidates about their positions on science-related issues. One of the questions was about the Bering Strait dam. Republican nominee Richard M. Nixon declined to answer the journal's questions; instead he submitted to them a lengthy policy statement of his view toward science. Democratic nominee John F. Kennedy answered the questions, saying the Siberia-Alaska dam was a technological undertaking that required close consideration, "but it is certainly worth exploring." Less than a week after saying this, Kennedy was elected president. During his short administration, the Bering Dam idea was never explored, at least outside the realm of classified government research.

But during that year another American became a zealous supporter of the Bering dam cause. Dr. Clyde Cowan was a physicist, renowned as the co-discoverer of the subatomic neutrino, and professor at Catholic University in Washington, D.C. After studying ideas for the dam, Cowan proposed a plan quite different from those of Markin and Borisov. He simply wanted the Bering Strait blocked up, and wrote that the job could be done with rock mined in Alaska, perhaps with the aid of nuclear explosions. No pumps would be employed to move Pacific water into the Arctic, or vice versa. All that was needed was a 56-mile-long barricade anchored on the Diomede Islands between Alaska and Siberia.

U.S. Weather Bureau experts analyzed what such a plain dam in the Bering Strait would do, and concluded it would "chill the Arctic Ocean slightly north of Alaska and Siberia." But Cowan said it would benefit these areas. By removing Arctic currents that now cool the Pacific Ocean, he wrote, the dam would gradually turn the whole Pacific basin into a "warm pond." Climate would warm up everywhere the ocean touched, including Alaska and Siberia. Rainfall would improve in the Soviet Union and throughout the western United States, where it could make wastelands arable and make the extraction of millions of barrels of shale oil economically feasible. Cowan told the *Los Angeles Times* his dam could turn Southern California, now technically a desert, into a "rain forest."

Dr. Cowan was convinced that control of the Bering Strait could act as a thermostat for global climate, that by regulating the amount of Arctic water that entered the Pacific basin, the nations controlling the dam could shape rainfall and temperature patterns throughout the planet. "The Bering Strait," he said, "is the only place I know where one could put his finger in the climate of the world and improve it."

In 1968 Richard M. Nixon was again preparing to run for the presidency, and Cowan traveled to New York in an effort to persuade him to endorse the Bering Dam idea. He nearly succeeded. At the direction of Nixon staff member Glenn Olds, and with Cowan's help, a speech was prepared in which candidate Nixon could praise imaginative science and technology. It read, in part:

> Millions of square miles of land in Mexico, in the United States—including Alaska—in Canada, in Russian Siberia, in Western China and in Korea, for instance, are either useless or are of such low rainfall that only the poorest existence may be eked out there.
>
> Too little rain over these vast areas of our planet. Too harsh a climate in untold tracts of otherwise bountiful land. I have outlined an area around the northern Pacific Ocean which—with a bit more rain per year—with more dependable rainfall each year—with a small rise in temperature over its northern parts—could support the entire population of the world today.
>
> Are these vast areas forever beyond our reach because of their climate? Must we hoard and divide and then divide again that little water which we find running in small streams across parched valleys and plains?
>
> Must we never hope to farm those great regions of the State of Alaska because the permanently frozen ground—the "permafrost"—lies so close to the surface—only a few inches down?
>
> Or can we—by taking thought—and then by taking prudent action as befits men in a modern world—make the desert bloom and let the rain fall where it is so sorely needed? . . .
>
> Suppose we used some of our great store of nuclear explosives to mine great quantities of rock in Alaska—and then used that rock to dam the Bering Strait? Suppose we closed off that narrow channel between Alaska and Siberia, and controlled the amount of cold Arctic water which may enter the Pacific Ocean? Would that turn the Pacific Ocean into a warm lake north of the equator? Would that provide rain and warmer climates to Northern China and Siberia, to Alaska

and Canada? Would it bring softer and lifegiving climate to
the deserts of our far west and to those of Mexico? . . .
Let us suppose for a moment that it might. And let us
imagine such a project underway. This would be one of the
greatest forces in history for the peaceful welding of vast
regions into one of common concern. . . . The Pacific
Community of Nations would become a community in more
than name—all dependent upon the safety and wise control
of a great dam which joins the continents. . . .
I am prepared for you, the scientists and engineers, to tell
me that the idea can not work for several . . . reasons. I am
not prepared—however—for *any* politician to tell me that it
cannot be considered for political reasons! That kind of talk
belongs to the administration which we hope to replace next
January.

Presumably for political reasons, Mr. Nixon never delivered this
speech. The next January he was inaugurated president, but during
his term in office never endorsed the Bering Dam idea publicly.

During President Nixon's re-election campaign in 1972,
Cowan joined two dozen other scientists in a Science and Engineering Council to support the president. Several council members were persuaded that a government study of the dam idea was
worthwhile, and Cowan prevailed on the president's brother Edward Nixon to bring his plan to White House attention.

Unknown to Dr. Cowan, the Bering Dam idea was at last
getting serious attention among policymakers, largely because
global conditions were changing. When the idea had been discussed decades earlier, Earth's climate still seemed to be warming. Why build a costly, controversial dam to do what nature
already was doing? But by 1971 the planet's climate was cooling
drastically. Experts blamed this for the drought that was starting to
cause mass famine in the African Sahel, and for changing weather
patterns everywhere. Some intelligence experts warned that the
cooling might mean a new Ice Age was beginning. If so, the world
could expect to perish by fire, not ice, for countries like the Soviet
Union and China would resort to nuclear war rather than see their
populations starve when food production fell because of the cold.
Only a madman would jump out of a third-floor window, some

said, unless the building was ablaze. The Bering Dam, which seemed a crazy way to enhance already-good climate during the 1950s, had by the 1970s become one of the best prospects for a technologic fix able to rescue the world from an Ice Age.

A crash program studying the dam's feasibility was secretly launched. In 1971 the United States and the Soviet Union laid plans for a series of experiments near the North Pole, the results of which would be used in evaluating the Bering Dam and other related ideas. The first of these was a joint U.S.–Soviet study of conditions in the Bering Sea, begun in 1972. The second was POLEX, the Polar Experiment of the Global Atmospheric Research Program. This basically consisted of the U.S. merging its research with an Arctic project the Soviets started in 1968. The third was a U.S.–Canadian study begun in 1973, AIDJEX, the Arctic Ice Dynamics Joint Experiment, which invited the Soviets to share data. Three Department of Defense agencies have been involved in AIDJEX. The computers of the Pentagon's climate analysis project, Nile Blue, were put to work studying the global cooling, the climatic impact of changing the Arctic ice pack, and ways the Arctic ice would be removed. A computer model was developed "to look at what sort of thing would affect the climate of North America," Rear Admiral Thomas D. Davies, a weapons expert for the U.S. Arms Control and Disarmament Agency, told a congressional subcommittee in 1974, "and it led to the conclusion that altering the Arctic icecap was the most effective thing you could do." Two years earlier the Defense Department had assured a Senate subcommittee that it had "not engaged in any experiments or field trials to melt the polar icecap."

In November 1974 President Ford and Soviet leaders met for a Summit Conference in Vladisvostok, Siberia. They secretly discussed the Northern Hemisphere's cooling trend, joint U.S.–Soviet research into it, and various ideas for halting the cooling with climate-modification technologies if the trend continues. The possibility of building a Bering Strait dam was touched upon, but deemed too controversial to mention publicly unless the cooling got significantly worse. But nations agreed to continue sharing research information about the cooling and ideas about ways to stop it.

Modifications of the Bering Dam idea have been considered. In

1958 civil engineer Gail Hathaway had suggested building "a gigantic dam stretching along the underwater ridge from Norway to Greenland." Its top would have to be about 200 feet above the water surface, said Hathaway, and it could be built either of concrete or, with the help of a long net of refrigeration coils, of ice created and maintained by freezing sea water. Power stations using tide and wind motion could run the refrigeration units. Like the Bering Dam, Hathaway's dam would control the flow of warm water into the Arctic—and much more water moves to and from the Arctic in the Atlantic than in the Pacific Ocean. Soviet engineer V. N. Stepenov had an alternative idea in 1963: use underwater nuclear explosives to blast away the submarine ridge southwest of the Faroe Islands. This would let cold water drain from the North Sea and help keep the Arctic open to the Atlantic Ocean even if sea levels fell. Both these ideas were rejected because they involved the politics of too many European nations. The Bering Dam would be smaller, simpler, and more nearly limited to a U.S.–Soviet agreement.

By comparision, other schemes for changing Earth's climate seemed crude or bizarre. Atomic reactors could be set up solely to heat the oceans, and metallic dust or CO_2 could be dumped into the air deliberately to enhance the atmosphere's "greenhouse effect" and hence warm the planet. But once done, these things could not be undone. A wrong calculation could mean that the world would fry rather than freeze. A Bering Dam, on the other hand, could at least be torn down again if it proved ill-conceived. At most, it risked melting Arctic ice that some experts said would not naturally refreeze. Some military planners talked of putting giant mirrors or unshielded nuclear reactors in orbit around the planet to act as "new suns." Such ideas were dismissed out of hand, as was the half-serious 1958 suggestion by Dr. Harry Wexler to detonate ten or more "clean" hydrogen bombs over the Arctic ice pack, using their heat and the vapor cloud they would cause to melt the Arctic Ocean.

In 1960 Soviet engineer Valentin B. Cherenkov proposed giving Earth a "ring" like that around Saturn. The ring would be made up of potassium dust, dispersed by "thousands" of rocket freighters 1500 kilometers above the planet. This dust ring was supposed to create "perpetual summer" and, in some regions, perpetual daylight in the higher latitudes by reflecting more sunlight into the Earth. This, he said, would create superb worldwide climate and make unlimited food growing possible. Critics of the scheme warned of ecological problems, especially the danger of melting polar ice. Some green plants grew wonderously during long summer days in Alaska and Siberia, experts said, but most fruits and vegetables would die in unending sunlight. The idea was abandoned.

Crazy as Cherenkov's plan seems, the Soviets could cite American precedents if they wanted to try it. In May 1963, over strong objections from scientists, the United States carried out Project Westford, the launching into high orbit of millions of copper filaments. These filaments spread into a thinly-spaced belt around our planet. The Department of Defense wanted this belt as a failsafe global reflector for military radio communications. Scientists objected that it would have unknown ecological consequences and probably would impair radio and optical astronomy. The military wish prevailed.

Likewise in July 1962 the Pentagon set off Project Starfish, a large nuclear device detonated at 250 miles altitude. Among other things, this created a new, perhaps permanent, band of trapped electrons in the force lines of Earth's magnetosphere. In April 1964, the U.S. tried to orbit a satellite with a plutonium power plant inside; it burned up over the Indian Ocean, spreading highly toxic radioactivity in the upper atmosphere that has been tracked over the globe, as atmospheric scientist Reginald Newell writes. Cherenkov's fantasy was scarcely more crazy than what Americans have already done in fact.

As Dr. Wexler noted, without a clear theory explaining why climate changes, human attempts to change it are risky. If we accepted the outmoded Ewing-Donn theory of why Ice Ages happen, we would see disaster in the two prime Soviet proposals to dam the Bering Strait and pump ocean currents through it as they wish. These schemes are designed to thaw the Arctic. But while they might succeed in melting the Arctic ice pack, they could also start a cycle of increased evaporation and precipitation that would cause the next Ice Age to come all the faster. By comparison, Dr. Cowan's plan seems almost sensible: by simply blocking the Bering Strait, the Arctic would be cooled even more, and this might reverse the present global cooling born of the shrinking ice pack there. As the Pacific Basin warmed, North America and eastern Siberia would warm too, so on balance Cowan's dam could improve Soviet climate even as it froze Russia's northern coasts. Also on balance, however, we must assume the Soviets would be unenthusiastic about such a dam. (The many other theories of climatic change would, of course, provide many other evaluations of various schemes to change world climate, including the Bering Dam ideas.)

If such a dam were built, asked Russian expatriate Dr. Serge Levitsky in 1959, who would control it? "In the hands of men with dreams of world conquest, the Bering Strait project could be used as a club to beat other nations, even whole continents, into submission." For this reason we can expect the United Nations to resist such an attempt at global climate modification too, and many nations to denounce such a project as an act of war. But if the global cooling continues, more than three-quarters of humankind will be threatened with cold, storminess, flood, or drought because of it. At some point in the cooling, a world consensus may be reached that something on the scale of the Bering Dam is worth risking. Given the selfish and limited vision of most nations, this agreement will only come when world temperatures have fallen 2°C. or more and the cooling has great momentum. By then all the cheap and comparatively safe methods of modifying world climate will be useless, and a cry will go up that more desperate measures be tried.

Sometimes people in burning buildings jump from windows twenty stories high. What people would do if they found themselves trapped in a freezing enclosure called "Spaceship Earth" remains to be seen.

15

Ice or Dice

How should we respond to Earth's changing climate? This is the most important political and practical question humankind now faces.

But it is a philosophical question too, for the answer we choose will come out of our world view and our values.

In America a century ago, as in the Soviet Union fifty years ago, such a decision would have been made easily. Man dominated the earth by divine right, and people believed that with science all things were possible. If nature caused a problem, we could invent new ways to manipulate nature for our benefit. We had stolen fire from the gods to warm ourselves, and someday when everything burnable on our small planet was gone we would split the atom for energy or fly away to other planets. We could climb into the heavens on manmade ladders of dream and invention.

Many still defend this view today, often by the reasoning of George Bernard Shaw: the reasonable person fits himself to the world, while the unreasonable person tries to bend the world to fit himself. Thus, said Shaw, all human progress has been achieved by unreasonable people. Every cure for disease or injury, every bit of cultivated food to nourish an expanding human race, every invention that lifted humankind from fear and ignorance and slavery, and even the human footprints on the moon, exist because some people refused to accept the world as they found it.

From this viewpoint the only question raised by the global cooling is how best can we use technology to control it; that we *should* control it is obvious. Typical of those holding this view is British scientist T. F. Gaskell, who in 1970 wrote, ''It would probably suit the human race best to put a brake on these natural

233

cycles [of warming and cooling] and perpetuate a comparatively steady set of climatic conditions."

I found these words in the public library copy of Gaskell's *Physics of the Earth*. They caught my eye because a reader had marked brackets around them, and in the book's margin written: "This guy is————!" (*expletive deleted*)

A growing number of people think the old view is wrong. Humankind is *in* nature, they believe, and not at war with it or divinely entitled to dominate it. Life is one vast ecosystem, and the destruction of the smallest species in the most faroff land threatens the web of life by which we all hang onto our shared spaceship, our ball of rock hurtling through the galaxy.

"Whenever someone sets out to remold the world, experience teaches that he is bound to fail," the Chinese philosopher Lao-tzu, founder of Taoism, wrote 2,500 years ago, "for nature is already as good as it can be. It cannot be improved upon. He who tries to redesign it, spoils it. He who tries to redirect it, misleads it." Many Occidentals today share this Oriental view.

For more than a century "progress" has been the most important cliché of all industrialized nations, regardless of ideology. But whether the accomplishments of science and industry merit being called progress, say ecologists, is much in doubt; more problems have been created than solved. A world that once feared the Black Plague now fears nuclear war, birth defects from chemical pollution, undrinkable water, and unbreathable air. A world that once feared famine now fears famine involving billions of humans rather than millions. Increasingly it seems that the accomplishments of humankind's invention have been based not on honest labor, but theft. We have broken into the places where natural energy was stored and stolen it for our own greedy desires. Ages ago humans were few, and our crime was petty theft, but now we are many, and our tampering with the delicate balances of nature can cause major dislocations in the global energy system. We have graduated to grand theft, and many people intuitively and logically conclude that some great natural law is about to catch up with us. Punishment may come quickly, by war or plague or famine, or more slowly, in the rising of the oceans or the coming of yet another Ice Age. A few see in such catastrophes the just hand of divine judgment and retribution against materialist sinners who refused to "return to nature."

The new and old view both have strengths and weaknesses. Those who cry "repent!" are right to warn that we have recklessly robbed natural storehouses and that resources like land, fresh water, and oil will soon be exhausted unless we mend our ways and become conservationists. They are wrong to say we must "return to nature," for we never left. The human-built skyscraper is as valid a "natural" structure as the bird's nest or the ant's anthill, if, as they say, we are part of nature. The issue instead is whether our present methods of survival improve our chances of going on for another million years or hinder them. Our species thinks and invents. This has made us the most flexible and durable of creatures, able to survive in wet lands and dry, in hot climate and cold, able both to fly and to travel under water. We who are descended from killers and dreamers and lovers and thieves cannot relinquish our heritage. Shaw was right about that.

But of course most of our experiments during the past million years failed. We are descended from the winners, the few human beings whose inventions succeeded in helping them adapt to a changing environment. Of these winners, most were eventually destroyed by their success: when conditions changed again, they were unable or unwilling to give up the old, once-successful way they had lived. The early agricultural civilizations fell because they learned how to irrigate land but not how to restore vital elements in the soil leeched by single crops and irrigation. After a few decades their fields lost fertility, much as fields in the United States overdosed with chemical fertilizer are doing today.

Our species has triumphed where we have relied on the "Noah Principle," the recognition that we must keep at least two of every kind alive on our spaceship Earth. The more variety we have, the better our ability to adapt to change. This, as Paul Ehrlich points out, is nature's way of doing things. We have failed, in most instances, when we became narrowminded and restricted our options. Perhaps the most frightening aspect of our modern technologies is their pervasiveness. Industrialization, once confined to Europe and the United States, is now spreading like a virus throughout the world. Hybrid seedgrains, once used in only a few places, are now grown around the planet. Such uniformity is dangerous. A new fungus preying on the genetic weakness of one hybrid grain could wipe out much of Earth's food supply. Harder to perceive, a standardization of machinery encourages a standardization of global thought, which reduces the pool of creativity from which new ideas can come in our species.

In the name of Noah we should respect both those who offer us ecological awareness and those who offer invention and imagination. Faced with global climatic instability—we know not whether its cause is human environmental tampering—we should evaluate our options in light of both perspectives. What follows is one attempt to do this:

Choice One: We can take no action. A handful of scientists denied evidence that Earth's climate was cooling until the 1970s, when bizarre weather throughout the world forced them to reconsider their views. By 1975 some of these same scientists—pointing to the Sahel, where some rain had returned after eight years; to the North Atlantic, where a change in global winds had brought slight warming between 1971 and 1974; and to rising temperatures in the Southern Hemisphere—were saying "the cooling is ended."

They could, of course, be right and if they are, any action to deal with the cooling seems foolish. But at least three factors must be weighed against their comforting belief:

1. On balance, evidence and theory suggest a serious risk of continued cooling, as the 1975 report of the National Academy of Sciences warned. Evidence to the contrary is, at best, debatable. As University of Wisconsin climatologist Reid Bryson says, "I never cease to be amazed how some people can deny drought exists when a land is without any rain for a decade, but if one cloud drops one inch of rain they cry 'the drought is broken.' " Those few who say "the cooling is ended" refused to call it a valid change in climate during the 1960s because world temperatures had been dropping for less than three decades, the span used by the World Meteorological Organization to distinguish climate from weather patterns. Then the cooling continued past three decades and had to be accepted as a shift in Northern Hemispheric climate. But now these same reluctant scientists declare that a three-year halt in the cooling in one small part of one ocean, and a few seasons of spotty rain in the drought-plagued regions of North Africa,

proves that world climate is warming. They may be right, but by their own standard of evidence we must wait until at least 1992 to agree, when another thirty years have passed. Between now and then, says Bryson, even if the cooling trend is being reversed by warming, the changeover will take decades; until at least the year 2,000 weather will probably be stormier, temperatures will likely be more extreme, monsoon rains will probably be less dependable, and oceans will remain colder than was the case in the first half of the century.

2. Even if warming is in the wings, its entrance spells trouble, for it would signal a period of severe unstable climate. This would entail most of the bad consequences of continued cooling—most notably disruption of global food production, and consequent famines. Given the potential pendulum effects of natural feedback mechanisms, a stable climate is the least likely outcome we could expect.

3. Suppose we assume, as did weather scientists interviewed by writer Nigel Calder, that the changes of continued cooling and of an Ice Age dawning within a century are one in ten, odds likened by one scientist to Russian roulette. The odds are in our favor, but consider the stakes being wagered: if the cooling continues, we can reasonably calculate that potentially two *billion* people could starve to death or die of other symptoms of chronic malnutrition by the year 2050. Potentially, we could all die if global famines and embargos on scarce resources, both caused by the cooling, lead to a world war.

We simply cannot afford to gamble against this possibility by ignoring it. We cannot risk inaction. Those scientists who say we should ignore the evidence and the theories suggesting Earth is entering a period of climatic instability are acting irresponsibly. The indications that our climate can soon change for the worse are too strong to be reasonably ignored.

Choice Two: We Can Adapt Ourselves to the Changing Climate. This choice is particularly sensible if what humankind faces is a "Little Ice Age" of the sort that chilled Europe and North America between 1350 and 1850. Ecologically, this is the wisest thing we can do, the course of action least likely to disrupt nature (that is, except the choice of doing nothing). Its prime problems are cost and feasibility. Already the less developed nations find it impossible to expand food production fast enough to meet rising populations, and industrialized nations find it increasingly difficult to balance their economies between inflation and recession. Adapting to a colder climate in the north latitudes, and to a drier climate nearer the equator, will require vast resources and almost unlimited energy. The cost over decades would run into *trillions* of dollars. This would impoverish the industrialized nations, and probably force all to resort to quasi-totalitarian systems of absolute food, water, and energy rationing and mandatory birth control. The poorest nations, if they survived at all, would become vassal states of nations with exportable food and energy. A few countries, such as equatorial Brazil, Zaire, and Indonesia, could emerge as climate-created superpowers. If the present warming in the Southern Hemisphere continues, Australia, South Africa, and Argentina would grow in world power too, as population centers and food-exporting regions.

If we are at the dawn of another Ice Age—a possibility that cannot be ruled out—then all attempts to adapt to the new climate while maintaining Earth's present population will fail. Famine will come, and only those who live in warmer parts of the planet or who can afford the high price of producing artificial climates in domed cities will survive. Mass migrations will begin, and wars will come as invaded tropical nations fight to stop the human influx. If this happens, global war is likely.

The cost of adapting the world to variable climate is high. The chance of such adaptation succeeding is only fair, and varies widely from nation to nation.

Choice Three: We can change weather and climate where we live. This can certainly help relieve symptoms of the global cooling, and could even reduce the potential harm of a Great Ice Age. But weather modification is as yet a limited tool. A 10 percent enhancement of rainfall is insufficient in regions where the shift in global climate has already reduced rainfall by 80 percent. Climate modification, even on a small scale, is ecologically dangerous because its full impact cannot be anticipated. A dam across Gibraltar Strait might, through a series of subtle natural feedback mechanisms, hasten the freezing of Northern Europe. Soviet diversion of rivers to create inland seas might cause drastic climate changes in the Arctic Ocean. Neighboring nations might look upon each other's environmental modification as acts of war and respond with other weapons.

One useful approach would be weather and climate modification under the auspices of the United Nations or some other international organization whose acts would be seen as in the interests of humankind rather than to benefit particular nations. The draft treaty banning environmental warfare produced in 1975 by the United States and Soviet Union opens the way for future international regulation of all weather and climate modification, but this achievement is unlikely in the twentieth century.

If successful, weather- and regional climate-modification techniques on a limited scale would be an inexpensive means of dealing with global cooling for the next several decades. If the cooling continues and intensifies, however, these tools will have little value; they will be overwhelmed. The size of regions warmed by reservoirs or atomic heating facilities or other such methods will shrink, and the snow will advance. The worth of cloud and storm seeding will diminish compared to the immensity of what must be done. For a few nations these technologies could make the difference between self-sufficiency and famine.

Some serious risks attend the use of limited weather and climate modification. The gains they can produce are marginal, and are probably of more value in warm climate than in cold.

Choice Four: We can try to control global climate. To ecologists this possibility is nightmarish because it involves the risk of environmental tampering on a world scale. Any attempt at such modification of world climate, such as blackening the polar ice caps or damming the Bering Strait, "would be highly irresponsible . . . until the consequences could be predicted better than is now possible," warned scientists taking part in the 1971 Report of the Study of Man's Impact on Climate (SMIC) sponsored by the Massachusetts Institute of Technology.

The problem is, how much must we know before we can take action? Assuming we will never know everything, what degree of certainty (and uncertainty) makes such a risk acceptable? Presumably such methods can be tried when and if the risk of inaction exceeds the risk of action. Some experts believe we are fast approaching that point. As Stephen H. Schneider of the National Center for Atmospheric Research recently said, the United States must soon begin planning its energy, food, aid, and other policy requirements *as if* global climate were to remain unstable, although we cannot prove that this will be the case. We can never gain certainty about the consequences of any method we could use to modify climate, says Schneider, so we should be willing to accept a lesser standard, plausibility, in judging whether to try such methods. (In his 1976 book *The Genesis Strategy*, Dr. Schneider proposes that, as Joseph advised Pharaoh in the Bible, we store up vast food reserves in years of good harvest to avert global famine in the bad years that surely will come as the climate deteriorates.)

Consider the idea of damming the Bering Strait. It has serious risks: its global impact on climate is debatable, and its political impact on the world community of nations is likely to be negative. If the idea is to have a chance to work, such a dam should probably be built soon, before the climatic cooling gathers enough momentum to become self-sustaining.

Suppose such a dam were built. At best it might be useless. At worst it could cause droughts from the Soviet Union to Brazil, while doing little to warm the world.

But it might succeed as its supporters have envisioned, and could forever after serve humankind as a thermostat for our planet, able to prevent future Ice Ages (perhaps even Ice Ages caused by small changes in the sun's radiation), and to open millions of acres of land to cultivation. By regulating the flow of heat into the Arctic Ocean it might regularize global weather, improving rainfall and banishing drought. Thus, even if the global cooling ended naturally, the Bering Dam could pay for itself overnight, for one year of severe bad weather can cost the U.S. economy alone between $6 and $20 billion. Every year the global cooling endures costs the world economy many times that figure, plus a large number of lives destroyed by famine.

As then-Senator John F. Kennedy said in 1960, the potential of a Bering Dam "is certainly worth exploring." If the global cooling continues, such exploration into ways we can deal with it becomes increasingly urgent. Ideas we once dismissed as unnecessary risks to our environment now deserve a second look.

Each of our four alternatives entails risk. What should we do? On one thing all agree: we desperately need as much new research information about weather and climate as we can get, and as quickly as possible. This will help us decide whether the cooling will continue, how fast our climate is changing, and why Earth's climate is increasingly unstable. It will suggest ways we can readjust the world weather machine that pumps heat poleward from the equator, and which aspects of that machine are too dangerous to tamper with.

A global weather experiment, like GATE but worldwide, had been planned for 1977. Because of a lack of funds and other political complications it has been postponed until 1978–79 and may be set back again unless money for its research is found. At present many scientific organizations are seeking support from Arab oil-producing nations and the Shah of Iran. Thus far the Shah's support promises to get the experiment off the ground. Results from this experiment will help us decide what to do about our changing climate.

But often scientists find only what they seek. Before GATE the first global weather experiment, BOMEX (the Barbados Oceanographic and Meteorological Experiment), was nearly a farce. Scientists set out merely to gather data, without any point in view, and compiled reams of incoherent information that computers were still trying to sort out years later. GATE was better, a directed effort to understand how tropical storms and cloud patterns form. What the World Meteorological Organization defines as the object of its next experiment will shape the results.

The 1978–79 experiment should concentrate its efforts on the nature and causes of Earth's changing climate, and on the workability of proposals to influence our planet's climate. In February 1976 the National Aeronautics and Space Administration (NASA) recommended that a large share ($4.6 billion) of its budget for the next several decades be spent on satellites to study and detect climatic changes; hopefully NASA will study climate manipulation too. The research of individual nations should also be directed to these concerns. In countries where citizens elect their leaders and lawmakers, the cooling should be made a political issue. Our planet's climatic cooling has already touched each of our lives in the higher prices we pay for food, fuel, and other resources, and in the future we can expect for our children and our world. We each deserve a voice in deciding what to do about it.

Before we can respond intelligently to the possibility of changing climate, a certain amount of scientific cynicism and political cowardice must be overcome.

Some weather scientists believe that it is inherently impossible to predict how climate will change or what consequences will come from any particular effort to modify climate. They prefer inaction "until more is known," but they would never agree that enough is known to warrant action. Their way of thinking is summed up by one of their favorite expressions: "We can't tell you what the weather will be like next Tuesday, so how can we predict what climate will be like in 30 years?"

While partly true—the art of computer-modelling weather and climate *is* primitive because funds for such research have been lacking—such reasoning is slightly flawed and deceptive. As Aldous Huxley used to joke, why is it that no fortune tellers are rich and no insurance companies ever go broke? The reason, said Huxley, is that fortune tellers deal in the specific: Will John Doe have an auto accident next Tuesday? Insurance companies, on the other hand, deal in statistical reality and generalities: How many auto accidents will occur in California during the coming decade? A similar distinction separates meteorologists from climatologists. A tiny change in wind or temperature can deflect the rainstorm the local weatherman predicted for Tuesday morning. But when a climatologist sees a major drop in hemispheric air temperatures, a large change in global air mass patterns, a huge drop in ocean temperatures, he can say with a high degree of statistical certainty that weather in general will take a turn for the worse—and he will be right, even though weather may seem balmy next Tuesday or even next year. (Speaking of insurance, notes Nigel Calder, the one-in-ten chance that an Ice Age may begin within a century "actuarially . . . means that the threat of ice reduces the life expectancy of everyone on the planet by several years.") The energy fabric on which daily weather in the Northern Hemisphere is painted has been severely distorted since 1950, and this will have consequences from now until year 2000 even if the cooling ceases today.

We can say with high probability today that the global monsoon rainfall will be below average for the remainder of this century. One human in four on our planet depends on those rains for survival. By the end of 1975 world food reserves remained below 30 days, despite good harvests in the monsoon belt and the United States (but poor harvests in Western Europe and disastrous ones in the Soviet Union). Experts estimated that as of early 1976 at least 270 million more people were alive on Earth than could be fed from that year's agricultural production. When the reserves ran out these people would die in a quiet famine—hidden in the death statistics in the countrysides of a hundred nations because, as Dr. Reid Bryson says, "Every country feeds its cities at the expense of the countryside—people there don't riot." In many parts of India, he notes, the death rate has doubled in recent years.

But human population grows, and to feed it more and more marginal land is pushed into production. This marginal land is often exquisitely sensitive to tiny changes in climate, so the human food supply becomes more and more vulnerable as time and the cooling continue.

In 1975 the National Academy of Sciences warned that the warm climate of our century has been "highly abnormal" and that we conceivably may be "on the brink of a [10,000 year] period of cold climate." NAS asked the government to more than triple its funding for climate research and experimentation, from $18 to $67 million per year, and to give appropriate consideration to actions that could deal with climatic change. "We simply cannot afford to be unprepared for . . . climatic catastrophe," said the NAS panel of prominent weather scientists.

In 1974 a group of the world's leading climatologists met in Bonn, Germany, and from their meeting came the following statement:

> The facts of the present climate change are such that the most optimistic experts would assign near certainty to major crop failures within a decade. If national and international

policies do not take these near-certain failures into account, they will result in mass deaths by starvation and probably in anarchy and violence that could exact a still more terrible toll. . . .

In 1975 most of these same scientists met in Berlin, in the wake of controversy about their earlier statement, and reaffirmed that all the best evidence available pointed to climatic changes and, in consequence, major crop failures "during the next decade."

Despite all available evidence, and all the warnings from experts, and the obvious peril of a world rushing to acquire nuclear weapons as famine nears, our political leaders are reluctant to act. Like children they wait for the toothache to go away and take comfort in those who advise that nothing need be done. But part of maturity is the willingness to face problems squarely, openly, honestly, and if our world is to survive we must demand such maturity from our leaders. We cannot continue as if nothing were wrong. To do so might prove both homicidal and, eventually, suicidal.

The dilemma we face in the global cooling was well expressed by the National Science Foundation in 1966. "Living things are adapted to the weather that actually prevails," it found, "and any change in that weather will be generally deleterious to them." This statement was part of a report on the hazards of weather and climate modification, but it applies equally to the global cooling nature has thrust upon us. "It seems reasonable that many organisms may be living at the edge of their capability to protect themselves," warned Stanford biologist Kendric Smith in 1975. "There might well be a key link in the plant-animal food chain on the brink right now—we just don't know."

But we *do* know that climate, the "weather that actually prevails" over decades, *is* changing, and we must assume this can have disastrous consequences. Which is the worse risk, to let the changes continue and pray they end soon, or to gamble that we can use technology to stabilize or reverse climate? The wisest course is a flexible mix of options: lay up what food we can and establish a global system to prevent famine, plus develop as many technologies as possible to help us adapt to changing climate, plus engage in carefully-controlled weather modification, plus provide massive funding to carry out the studies needed to assess the likelihood of continued climatic changes and the best means to modify climate if that proves necessary.

Such a program entails enormous costs and some risks—the risk, for example, that food hoarding will increase if political leaders speak openly of the dangers of changing climate or of the cooling continuing. But what other choices do we have?

Appendix One

THE SHAPE OF COOLING

In January 1975 the National Academy of Sciences issued a report entitled *Understanding Climatic Change: A Program for Action*. There is, it said, "a finite possibility that a serious worldwide cooling could befall the earth within the next hundred years."

From such a staid group the statement was surprisingly strong, but the panel of experts authorizing it had good reasons for what they said: (1) global climate was already cooling; Northern Hemispheric temperatures had been in steady decline since the 1940s; (2) the period of warm climate Earth enjoyed between 1880 and 1940 was highly abnormal; when considered in the context of world history, we have seen the warmest century of the last millennium, which was part of the warmest 10,000-year period of the last million years, and this odd warmth cannot be expected to last; (3) climate changes in the past have followed well-defined cycles, and if these cycles continue we can anticipate colder climate to return soon.

Two of the cycles run at intervals of 20,000 years and 2,500 years during interglacials (periods of ice retreat). The shorter cycle apparently reached its coolest point during the recent "Little Ice Age" from 1430 until 1850, and is now moving toward increased warmth. But the longer cycle brings severe cooling every 10,000 years, and it last did this 10,000 years ago. Thus, said the NAS report, "the question naturally arises as to whether we are indeed on the brink of a [10,000-year] period of colder climate."

The NAS provided a series of charts showing cycles in climate and global ice volume. The regularity of cycles lasting 100,000 years, 20,000 years, and 2,500 years is evident:

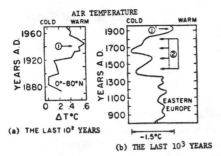

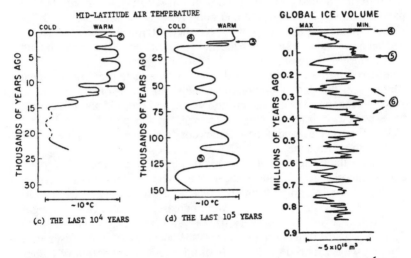

LEGEND

1. THERMAL MAXIMUM OF 1940s

2 LITTLE ICE AGE

3. YOUNGER DRYAS COLD INTERVAL

4. PRESENT INTERGLACIAL (HOLOCENE)

5. LAST PREVIOUS INTERGLACIAL (EEMIAN)

6. EARLIER PLEISTOCENE INTERGLACIALS

FIGURE A-1

Generalized trends in global climate: the past million years. (a) Changes in the five-year average surface temperatures over the region 0-80°N during the last 100 years (Mitchell, 1963). (b) Winter severity index for eastern Europe during the last 1000 years (Lamb, 1969). (c) Generalized midlatitude northern hemisphere air-temperature trends during the last 15,000 years, based on changes in tree lines (LaMarche, 1974), marginal fluctuations in alpine and continental glaciers (Denton and Karlén, 1973), and shifts in vegetation patterns recorded in pollen spectra (van der Hammen et. al., 1971). (d) Generalized northern hemisphere air-temperature trends during the last 100,000 years, based on midlatitude sea-surface temperature and pollen records and on worldwide sea-level records. (e) Fluctuations in global ice-volume during the last 1,000,000 years as recorded by changes in isotopic composition of fossil plankton in deep-sea core V28-238 (Shackleton and Opdyke, 1973). See legend for identification of symbols (1) through (6).

Appendix Two

THE U.S. SENATE RESOLUTION ON ENVIRONMENTAL WARFARE

On July 11, 1973—more than a year after Department of Defense officials announced that weather modification had been used as a weapon of war in Southeast Asia by the United States—the U.S. Senate passed a resolution urging a ban on all military *research and* use of environmental modification. It was introduced by Senator Claiborne Pell (D.-Rhode Island) and co-authored by eighteen other Senators. The Senate voted 82 to 10 to approve the resolution.

Such resolutions are said to express the "mood" of the United States Senate, and as such may influence policy. But they lack the force of law.

93D CONGRESS   1ST SESSION

S. RES. 71
[Report No. 93–270]

IN THE SENATE OF THE UNITED STATES

FEBRUARY 22, 1973

MR. PELL (for himself, MR. BAYH, MR. CASE, MR. CHURCH, MR. CRANSTON, MR. GRAVEL, MR. HART, MR. HOLLINGS, MR. HUGHES, MR. HUMPHREY, MR. JAVITS, MR. KENNEDY, MR. MCGOVERN, MR. MONDALE, MR. MUSKIE, MR. NELSON, MR. STEVENSON, MR. TUNNEY, and MR. WILLIAMS) submitted the following resolution; which was referred to the Committee on Foreign Relations

JUNE 27 (legislative day, JUNE 25), 1973

Reported by MR. PELL, with amendments

JULY 11, 1973

Considered, amended, and agreed to

RESOLUTION

Expressing the sense of the Senate that the United States Government should seek the agreement of other governments to a proposed treaty prohibiting the use of any environmental or geophysical modification activity as a weapon of war, or the carrying out of any research or experimentation directed thereto.

Whereas there is vast scientific potential for human betterment through environmental and geophysical controls; and

Whereas there is great danger to the world ecological system if environmental and geophysical modification activities are not controlled or if used indiscriminately; and

Whereas the development of weapons-oriented environmental and geophysical modification activities will create a threat to peace and world order; and

Whereas the United States Government should seek agreement with other governments on the complete cessation of any research, experimentation, or use of any such activity as a weapon of war: Now, therefore, be it

Resolved, That it is the sense of the Senate that the United States Government should seek the agreement of other governments, including all Permanent Members of the Security Council of the United Nations, to a treaty along the following general lines which will provide for the complete cessation of any research, experimentation, and use of any environmental or geophysical modification activity as a weapon of war:

"The Parties to this Treaty,

"Recognizing the vast scientific potential for human betterment through environmental and geophysical controls,

"Aware of the great danger to the world ecological system of uncontrolled and indiscriminate use of environmental and geophysical modification activities,

"Recognizing that the development of weapons-oriented environmental and geophysical modification techniques will create a threat to peace and world order,

"Proclaiming as their principal aim the achievement of an agreement on the complete cessation of research, experimentation, and use of environmental and geophysical modification activities as weapons of war,

"Have agreed as follows:

"ARTICLE I
"(1) The States Parties to this Treaty undertake to prohibit and prevent, at any place, any environmental or geophysical modification activity as a weapon of war;

"(2) The prohibition in paragraph 1 of this article shall also apply to any research or experimentation directed to the development of any such activity as a weapon of war, but shall not apply to any research, experimentation, or use for peaceful purposes;

"(3) The States Parties to this Treaty undertake not to assist, encourage or induce any State to carry out activities referred to in paragraph 1 of this article and not to participate in any other way in such actions.

"ARTICLE II

"In this Treaty, the term 'environmental or geophysical modification activity' includes any of the following activities:

"(1) any weather modification activity which has as a purpose, or has as one of its principal effects, a change in the atmospheric conditions over any part of the earth's surface, including, but not limited to, any activity designed to increase or decrease precipitation, increase or suppress hail, lightning, or fog, and direct or divert storm systems;

"(2) any climate modification activity which has as a purpose, or has as one of its principal effects, a change in the long-term atmospheric conditions over any part of the earth's surface;

"(3) any earthquake modification activity which has as a purpose, or has as one of its principal effects, the release of the strain energy instability within the solid rock layers beneath the earth's crust;

"(4) any ocean modification activity which has as a purpose, or has as one of its principal effects, a change in the ocean currents or the creation of a seismic disturbance of the ocean (tidal wave).

"ARTICLE III

"Five years after the entry into force of this Treaty, a conference of Parties shall be held at Geneva, Switzerland, in order to review the operation of this Treaty with a view to assuring that the purposes of the preamble and the provisions of the Treaty are being realized. Such review shall take into account any relevant technological developments in order to determine whether the definition in Article II should be amended.

"ARTICLE IV

"1. Any Party may propose an amendment to this Treaty. The text of any proposed amendment shall be submitted to the Depositary Governments which shall circulate it to all parties to this Treaty. Thereafter, if requested to do so by one-third or more of the Parties, the Depositary Governments shall convene a conference, to which they shall invite all the Parties, to consider such an amendment.

"2. Any amendment to this Treaty shall be approved by a majority of the votes of all the Parties to this Treaty. The amendment shall enter into force for all Parties upon the deposit of instruments of ratification by a majority of all the Parties.

"ARTICLE V

"1. This Treaty shall be of unlimited duration.

"2. Each Party shall, in exercising its national sovereignty, have the right to withdraw from the Treaty if it decides that extraordinary events, related to the subject matter of this Treaty, have jeopardized the supreme interests of its country. It shall give notice of such withdrawal to all other Parties to the Treaty three months in advance.

"ARTICLE VI

"1. This Treaty shall be open to all States for signature. Any State which does not sign this Treaty before its entry into force in accordance with paragraph 3 of this Article may accede to it at any time.

"2. This Treaty shall be subject to ratification by signatory States. Instruments of ratification and instruments of accession shall be deposited with the Governments of the United States of America, [and others] which are hereby designated the Depositary Governments.

"3. This Treaty shall enter into force after its ratification by the States, the Governments of which are designated Depositaries of the Treaty.

"4. For States whose instruments of ratification or accession are deposited subsequent to the entry into force of this Treaty, it shall enter into force on the date of the deposit of their instruments of ratification or accession.

"5. The Depositary Governments shall promptly inform all signatory and acceding States of the date of each signature, the date of deposit of each instrument of ratification of and accession to this Treaty, the date of its entry into force, and the date of receipt of any requests for conferences or other notices.

"6. This Treaty shall be registered by the Depositary Governments pursuant to Article 102 of the Charter of the United Nations."

Appendix Three

THE SOVIET UNITED NATIONS RESOLUTION ON ENVIRONMENTAL WARFARE

On September 24, 1974, the Soviet Union submitted a draft resolution to the United Nations General Assembly. It included a draft "Convention on the prohibition of action to influence the environment and climate for military and other purposes incompatible with the maintenance of international security, human well-being and health. . . ." This was revised and presented as a resolution for General Assembly acceptance on November 21, 1974, co-sponsored by Afghanistan, Argentina, Bangladesh, Barbados, Bulgaria, the Byelorussian Soviet Socialist Republic, Cuba, Czechoslovakia, Democratic Yemen, Egypt, Finland, the German Democratic Republic, Ghana, Hungary, India, Iraq, Kenya, Mauritius, Mongolia, Poland, the Syrian Arab Republic, the Ukrainian Soviet Socialist Republic, and the United Republic of Cameroon.

TEXT:
RECOMMENDATION OF THE FIRST COMMITTEE

7. The First Committee recommends to the General Assembly the adoption of the following draft resolution:

Prohibition of action to influence the environment and climate for military and other purposes incompatible with the maintenance of international security, human well-being and health

The General Assembly,

Noting the concern of peoples to consolidate peace and to pursue efforts designed to save mankind from the danger of using new means of warfare, to limit the arms race and to bring about disarmament,

Bearing in mind that, under conditions of continuous scientific and technological progress, new possibilities arise for using the results of this progress not only for peaceful but also for military purposes,

Convinced that the prohibition of action to influence the environment and climate for military and other hostile purposes, which are incompatible with the maintenance of international security, human well-being and health, would serve the cause of strengthening peace and averting the threat of war,

Taking into account the profound interest of States and peoples in the adoption of measures to preserve and improve the environment and to modify or moderate the climate solely for peaceful purposes for the benefit of present and future generations,

1. *Considers it necessary* to adopt, through the conclusion of an appropriate international convention, effective measures to prohibit action to influence the environment and climate for military and other hostile purposes, which are incompatible with the maintenance of international security, human well-being and health;

2. *Takes note* of the draft international "Convention on the prohibition of action to influence the environment and climate for military and other purposes incompatible with the maintenance of international security, human well-being and health" submitted to the General Assembly by the Union of Soviet Socialist Republics as well as other points of view and suggestions put forward during the discussion of this question;

3. *Requests* the Committee on Disarmament to proceed as soon as possible to achieving agreement on the text of such a convention and to submit a report on the results achieved for consideration by the General Assembly at its thirtieth session;

4. *Requests* the Secretary-General to transmit to the Committee on Disarmament all documents relating to the discussion by the General Assembly at its twenty-ninth session of the item entitled "Prohibition of action to influence the environment and climate for military and other purposes incompatible with the maintenance of international security, human well-being and health";

5. *Decides* to include in the provisional agenda of its thirtieth session an item entitled "Prohibition of action to influence the environment and climate for military and other hostile purposes, which are incompatible with the maintenance of international security, human well-being and health."

ANNEX

CONVENTION ON THE PROHIBITION OF ACTION TO INFLUENCE THE ENVIRON-
MENT AND CLIMATE FOR MILITARY AND OTHER PURPOSES INCOMPATIBLE WITH
THE MAINTENANCE OF INTERNATIONAL SECURITY, HUMAN WELL-BEING AND
HEALTH

Union of Soviet Socialist Republics: draft convention

The Stateş Parties to this Convention,

Guided by the interests of consolidating peace and wishing to contribute to the cause of saving mankind from the danger of using new means of warfare, limiting the arms race and bringing about disarmament,

Taking into account that, under conditions of continuous scientific and technological progress, new possibilities arise for using the result of this progress not only for peaceful but also for military purposes,

Considering that action to influence the environment and climate for military purposes may represent an exceptional danger to universal peace and security as well as to human well-being and health,

Expressing the profound interest of States and peoples in the adoption of measures to preserve and improve the environment for the benefit of present and future generations,

Desiring to contribute to the deepening of confidence among peoples and to the further improvement of the international situation,

Striving to co-operate in implementing the purposes and principles of the Charter of the United Nations,

Have agreed on the following:

ARTICLE I
Each of the parties to the Convention undertakes not to use meteorological, geophysical or any other scientific or technological means of influencing the environment, including the weather and climate, for military and other purposes incompatible with the maintenance of international security, human well-being and health, and, furthermore, never under any circumstances to resort to such means of influencing the environment and climate or to carry out preparations for their use.

ARTICLE II
1. For the purposes of this Convention, the activities referred to in article I of this Convention consist of those active influences on the surface of the land, the sea-bed and the ocean floor, the depths of the earth, the marine environment, the atmosphere or on any other elements of the environment that may cause damage by the following means:

(a) Introduction into the cloud systems (air masses) of chemical reagents for the purpose of causing precipitation (formation of clouds) and other means of bringing about a redistribution of water resources;

(b) Modification of the elements of the weather, climate and the hydrological system on land in any part of the surface of the earth;

(c) Direct or indirect action to influence the electrical processes in the atmosphere;

(d) Direct or indirect disturbance of the elements of the energy and water balance of meteorological phenomena (cyclones, anticyclones, cloud front systems);

(e) Direct or indirect modifications of the physical and chemical parameters of the seas and oceans, the seashore, sea-bed and ocean floor that may lead to a change in the hydrological system, water interchange process and the ecology of the biological resources of the seas and oceans;

(f) Direct or indirect stimulation of seismic waves by any methods or means that may produce earthquakes and accompanying processes and phenomena, or destructive ocean waves, including tsunami;

(g) Direct or indirect action on the surface of an area of water that may lead to a disturbance of the thermal and gaseous interchange between the hydrosphere and the atmosphere;

(h) The creation of artificial continuous electromagnetic and acoustic fields in the oceans and seas;

(i) Modification of the natural state of the rivers, lakes, swamps and other aqueous elements of the land by any methods or means, leading to reduction in the water-level, drying up, flooding, inundation, destruction of hydrotechnical installations or having other harmful consequences;

(j) Disturbance of the natural state of the lithosphere, including the land surface, by mechanical, physical or other means, causing erosion, a change in the mechanical structure, desiccation or flooding of the soil, or interference with irrigation or land improvement systems;

(k) The burning of vegetation and other actions leading to a disturbance of the ecology of the vegetable and animal kingdom;

(l) Direct or indirect action to influence the ionized or ozone layers in the atmosphere, the introduction of heat and radiant energy absorbing agents in the atmosphere and the contiguous layer, or other action that might lead to disturbances of the thermal and radiation equilibrium of the earth-atmosphere-sun system.

2. Subsequently, in accordance with the provisions of this Convention,

the list of actions enumerated in paragraph 1 of this article may be supplemented or amended depending upon the progress of scientific and technological research.

ARTICLE III
Each of the parties to the Convention undertakes to refrain from assisting, encouraging or inducing any State, group of States or international organizations whatsoever to carry out activities that violate the provisions of this Convention, as well as to refrain from participating either directly or indirectly in such activities carried out by other States or international organizations.

ARTICLE IV
Each party to this Convention undertakes, in accordance with its own constitutional procedures, to adopt the necessary measures to prohibit and prevent any activity carried out in violation of the provisions of this Convention anywhere whatsoever within its jurisdiction or under its control.

ARTICLE V
Nothing in this Convention shall impede the economic or scientific and technological development of the parties to the Convention or international economic and scientific co-operation in the utilization, preservation and improvement of the environment for peaceful purposes.

ARTICLE VI
Any party to this Convention that learns that any other party to the Convention is acting in violation of the obligations flowing from the provisions of the Convention may lodge a complaint with the Security Council of the United Nations. Such a complaint shall contain all possible evidence to support the grounds for the complaint, together with a request that it be considered by the Security Council.

Each party to this Convention undertakes to co-operate in carrying out any investigations that the Security Council may undertake in accordance with the provisions of the Charter of the United Nations on the basis of the complaint received by the Council. The Security Council shall inform the States parties to the Convention of the results of such investigations.

ARTICLE VII
Each party to this Convention undertakes to furnish or support assistance provided in accordance with the Charter of the United Nations to any party to the Convention that may make such a request, in the event that

the Security Council adopts a decision to the effect that that party has been subjected to danger as a result of the violation of the Convention.

ARTICLE VIII
Any party may propose an amendment to this Convention. Each proposed amendment shall be submitted to the depositary Governments and shall be transmitted by them to all parties to the Convention, which shall inform the depositary Governments of the adoption or rejection of the amendment at the earliest possible date after receiving it.

The amendment shall enter into force for each party accepting it after its adoption by the majority of parties to the Convention, including the depositary Governments, and subsequently for each remaining party on the day on which it adopts that amendment.

ARTICLE IX
Five years after the entry into force of this Convention, or before that date, if the majority of parties to the Convention so request by submitting a proposal for that purpose to the depositary Governments, a conference of States parties to the Convention shall be convened in . . . for the purpose of considering the operation of the Convention, in order to ensure that its provisions are being implemented. During such consideration, account shall be taken of all new scientific and technological achievements that may relate to the Convention.

ARTICLE X
This Convention shall be of a permanent nature.

Each party to this Convention shall have the right, within the context of the realization of its own State sovereignty, to secede from the Convention, if it decides that exceptional circumstances connected with the content of the Convention have threatened the supreme interests of its country. It shall notify all other States parties to the Convention and the Security Council of the United Nations three months prior to its secession. The notification shall contain an account of the exceptional circumstances which, in the view of that party, have threatened its supreme interests.

ARTICLE XI
1. This Convention shall be open for signature by all States. Any State that does not sign the Convention before its entry into force in accordance with paragraph 3 of this article may accede to it at any time.

2. This Convention shall be subject to ratification by the signatory States. The instruments of ratification and the documents of accession

shall be deposited with the Governments of . . ., which are hereby designated the depositary Governments.

3. This Convention shall enter into force after the deposit of the instruments of ratification by . . . Governments, including Governments designated the depositary Governments of the Convention.

4. For those States whose instruments of ratification or documents of accession are deposited after the entry into force of this Convention, it shall enter into force on the day of deposit of their instruments of ratification or documents of accession.

5. The depositary Governments shall forthwith notify all States that sign or accede to this Convention of the date of each signature, the date of deposit of each instrument of ratification or document of accession, the date of the entry into force of the Convention, and the receipt by them of other information.

This Convention shall be registered by the depositary Governments in accordance with Article 102 of the Charter of the United Nations.

ARTICLE XII
This Convention, of which the Chinese, English, French, Russian and Spanish texts are equally authentic, shall be deposited in the archives of the depositary Governments. Certified copies of the Convention shall duly be forwarded by the depositary Governments to the Governments of States signing the Convention or acceding to it.

> (Official English text, United Nations
> General Assembly document A/9910,
> 6 December 1974.)

On November 22, 1974, the draft resolution was adopted by a roll-call vote of 102 to none, with seven abstentions. The voting was as follows:

In favor: Algeria, Argentina, Australia, Austria, Bahrain, Belgium, Bhutan, Brazil, Bulgaria, Burma, Byelorussian Soviet Socialist Republic, Canada, Central African Republic, Chad, Colombia, Congo, Costa Rica, Cuba, Cyprus, Czechoslovakia, Democratic Yemen, Denmark, Dominican Republic, Ecuador, Egypt, El Salvador, Ethiopia, Fiji, Finland, Gambia, German Democratic Republic, Germany (Federal Republic of), Ghana, Greece, Guatemala, Guinea, Guyana, Hungary, India, Indonesia, Iran, Iraq, Ireland, Israel, Italy, Ivory Coast, Japan, Jordan, Kenya, Kuwait, Laos, Lebanon, Liberia, Libyan Arab Republic, Madagascar, Malaysia, Mauritania, Mexico, Mongolia, Morocco,

Nepal, Netherlands, New Zealand, Nicaragua, Niger, Nigeria, Norway, Oman, Pakistan, Peru, Philippines, Poland, Portugal, Qatar, Romania, Rwanda, Senegal, Sierra Leone, Singapore, Somalia, Spain, Sri Lanka, Sudan, Swaziland, Sweden, Syrian Arab Republic, Thailand, Tunisia, Turkey, Uganda, Ukrainian Soviet Socialist Republic, Union of Soviet Socialist Republics, United Arab Emirates, United Kingdom of Great Britain and Northern Ireland, United Republic of Cameroon, United Republic of Tanzania, Upper Volta, Venezuela, Yugoslavia, Zaire, Zambia, Afghanistan.

Against: None.

Abstaining: Burundi, Chile, France, Gabon, Mali, Paraguay, United States of America.

Appendix Four

THE SOVIET-AMERICAN JOINT DRAFT TREATY

On July 3, 1974, during a summit meeting, the United States and the Soviet Union issued the following joint statement:

The United States of America and the Union of Soviet Socialist Republics:

Desiring to limit the potential danger to mankind from possible new means of warfare;

Taking into consideration that scientific and technical advances in environmental fields, including climate modification, may open possibilities for using environmental modification techniques for military purposes;

Recognizing that such use could have widespread, long-lasting, and severe effects harmful to human welfare;

Recognizing also that proper utilization of scientific and technical advances could improve the inter-relationship of man and nature;

1. Advocate the most effective measures possible to overcome the dangers of the use of environmental modification techniques for military purposes.

2. Have decided to hold a meeting of United States and Soviet representatives this year for the purpose of exploring this problem.

3. Have decided to discuss also what steps might be taken to bring about the measures referred to in paragraph 1.

The first of these meetings was held in Moscow from November 1 through November 5, 1974. The second took place in Washington, D.C., between February 24 and March 5, 1975.

Out of these talks came a draft treaty on environmental modification, which was presented simultaneously by the United States and Soviet Union to the thirty-one-nation Geneva Disarmament Conference on August 21, 1975. The draft treaty's key provisions are as follows:

TEXT: *Convention on the Prohibition of Military or any other Hostile Use of Environmental Modification Techniques*

The States Party to this Convention,

Guided by the interest of consolidating peace, and wishing to contribute to the cause of limiting the arms race, and of bringing about disarmament, and of saving mankind from the danger of using new means of warfare;

Recognizing that scientific and technical advances may open new possibilities with respect to modification of the environment;

Realizing that military use of environmental modification techniques could have widespread, long-lasting or severe effects harmful to human welfare, but that the use of environmental modification techniques for peaceful purposes could improve the interrelationship of man and nature and contribute to the preservation and improvement of the environment for the benefit of present and future generations;

Desiring to limit the potential danger to mankind from means of warfare involving the use of environmental modification techniques;

Desiring also to contribute to the strengthening of trust among nations and to the further improvement of the international situation in accordance with the purposes and principles of the Charter of the United Nations,

Have agreed as follows:

ARTICLE I
1. Each State Party to this Convention undertakes not to engage in military or any other hostile use of environmental modification tech-

niques having widespread, long-lasting or severe effects as the means of destruction, damage or injury to another State Party.

2. Each State Party to this Convention undertakes not to assist, encourage or induce any State, group of States or international organization to engage in activities contrary to the provision of paragraph 1 of this Article.

ARTICLE II

As used in Article I, the term "environmental modification techniques" refers to any technique for changing—through the deliberate manipulation of natural processes—the dynamics, composition or structure of the Earth, including its biota, lithosphere, hydrosphere, and atmosphere, or of outer space, so as to cause such effects as earthquakes and tsunamis, an upset in the ecological balance of a region, or changes in weather patterns (clouds, precipitation, cyclones of various types and tornadic storms), in the state of the ozone layer or ionosphere, in climate patterns, or in ocean currents.

ARTICLE III

The provisions of this Convention shall not hinder the use of environmental modification techniques for peaceful purposes by States Party, or international economic and scientific co-operation in the utilization, preservation and improvement of the environment for peaceful purposes.

ARTICLE IV

Each State Party to this Convention undertakes, in accordance with its constitutional processes, to take any necessary measures to prohibit and prevent any activity in violation of the provisions of the Convention anywhere under its jurisdiction or control.

ARTICLE V

1. The States Party to this Convention undertake to consult one another and to cooperate in solving any problems which may arise in relation to the objectives of, or in the application of the provisions of this Convention. Consultation and cooperation pursuant to this Article may also be undertaken through appropriate international procedures within the framework of the United Nations and in accordance with its Charter.

2. Any State Party to this Convention which finds that any other State Party is acting in breach of obligations deriving from the provisions of the Convention may lodge a complaint with the Security Council of the United Nations. Such a complaint should include all possible evidence confirming its validity, as well as a request for its consideration by the Security Council.

3. Each State Party to this Convention undertakes to cooperate in carrying out any investigation which the Security Council may initiate, in accordance with the provisions of the Charter of the United Nations, on the basis of the complaint received by the Council. The Security Council shall inform the States Party to the Convention of the results of the investigation.

4. Each State Party to this Convention undertakes to provide or support assistance, in accordance with the United Nations Charter, to any Party to the Convention which so requests, if the Security Council decides that such Party has been harmed or is likely to be harmed as a result of violation of the Convention.

[Articles VI–IX deal with amendment, duration, ratification, and official translation of the treaty.]

(Text from the U.S. Arms Control and Disarmament Agency, Washington, D.C.)

Appendix Five

Three of the strongest critics of military study and use of environmental warfare techniques are Congressman Gilbert Gude (R.-Maryland), Donald M. Fraser (D.-Minnesota) and Senator Claiborne Pell (D.-Rhode Island).

Although they oppose all military use of weather and climate modification, these lawmakers have taken the lead in urging peaceful use of it. On April 23, 1975, they asked the president to maintain and expand research and programs dedicated to changing weather and climate for human benefit. Their request was made in a letter reading as follows:

GILBERT GUDE
8TH DISTRICT, MARYLAND

COMMITTEE ON
GOVERNMENT OPERATIONS
SUBCOMMITTEE:
CONSERVATION, ENERGY AND NATURAL
RESOURCES

COMMITTEE ON
DISTRICT OF COLUMBIA
RANKING MINORITY MEMBER

SELECT COMMITTEE ON AGING
SUBCOMMITTEE:
RETIREMENT INCOME AND EMPLOYMENT

GORDON L. HAWK
ADMINISTRATIVE ASSISTANT

WILLIAM A. REINSCH
ROBERTA J. AVANCENA
BRUCE W. WOOD
LEGISLATIVE ASSISTANTS

WILLIAM GRIGG
PRESS ASSISTANT

104 CANNON HOUSE OFFICE BUILDING
WASHINGTON, D.C. 20515
TELEPHONE: 225-5341

Congress of the United States
House of Representatives
Washington, D.C. 20515

April 23, 1975

The President
The White House

Dear Mr. President:

As authors of several resolutions for outlawing environmental modification as a weapon of war, we now write recommending government work in the peaceful uses of such modification that could help to promote energy conservation, safeguard the environment and stabilize agricultural production. In sending these recommendations, we wish to make clear that we support continued research, particularly into weather modification for peaceful purposes, regarding which we believe there currently exist numerous opportunities for its applications.

The role of weather modification in energy conservation was sharply outlined in a recent example which came to our attention. Comming from Boston to Washington, a recent flight was delayed by bad weather and according to one passenger's calculations, as much fuel was exhausted around Washington while the plane waited to land as was consumed during the entire flight from Boston. This is only one example of the energy costs of bad weather, but weather conditions being what they are, it is a frequent case. Research into fog dissipation is precisely the kind of work which can reduce those costs.

We are only beginning to research and understand how our own industrial development has inadvertently modified weather and environment. Studies are beginning to show differences in temperature and air quality over urban and industrial areas, which affect the immediate environment as well as influence weather downwind. There is sufficient growing suspicion that inadvertent environmental modification can help produce extremes of weather, such as drought, to warrant further investigation and research.

The implications of weather modification for agriculture are obvious and various efforts to enhance rainfall have been going on for years. These efforts, however, need coordination and careful study to help determine what approaches are productive, what types of weather formation are most susceptible to modification and how modification in one area affects weather elsewhere. Clearly, the potential for increased

The President
April 23, 1975
Page 2

agriculture output--both domestically and worldwide--is great.

Given these opportunities, it is unfortunate that civilian directed
research has been diffuse. The fiscal 1975 budget shows weather
modification projects in six agencies as follows:

	FY 1973	FY 1974	FY 1975
Department of Agriculture	366	270	150
Department of Commerce	4,779	4,673	4,575
Department of Defense	(1,209)	(1,161)	(1,300)
Army	160	96	
Navy	404	399	555
Air Force	645	666	745
Department of the Interior	6,370	3,900	3,455
Department of Transportation	1,067	1,397	1,520
National Science Foundation	5,790	4,000	4,270
TOTAL	19,581	15,401	15,270

and a division by function as follows:

	FY 1973	FY 1974	FY 1975	Agencies
Precipitation Modification	5,472	3,735	3,279	DOC-DOI
Fog and Cloud Modification	1,541	1,194	1,264	DOD-DOT
Hail Suppression	2,860	2,000	2,100	NSF
Lightning Modification	624	330	366	DOA-DOD-NSF
Hurricane and Severe Storm Modification	1,818	1,741	1,816	DOC
Social, Economic, Legal and Ecological Studies	1,740	1,310	1,110	DOI-NSF
Inadvertent Modification of Weather and Climate	3,252	3,643	4,398	DOC-DOT-NSF
Support and Services	2,274	1,475	937	DOC-DOI-NSF
TOTAL	19,581	15,401	15,270	

Although in some respects the National Oceanographic and Atmospheric
Administration gathers data on all these projects, it does not really function
as a lead agency or exert sufficient direction, coordination or control
over the civilian or military projects. It is clear from the second
chart, furthermore, that considerable overlap and possible duplication
exists. We believe, however, that in a field as diverse and speculative
as this, a greater degree of centralization is desirable. This same recommendation has been made on a number of occasions by the National Advisory
Committee on Oceans and Atmosphere:

The President
April 23, 1975
Page 3

NACOA finds that, although we appear to stand on the threshold
of practical weather modification, and some facets are operational,
in other applications a great deal of complex research still
needs to be done. Unless the scientific manpower and funding
are better directed, we assuredly will continue to make very
slow progress towards weather control. NACOA therefore reit-
erates its RECOMMENDATIONS of last year that:
    The many small programs in weather modification
    now scattered widely through the Federal agencies
    be focused and coordinated under NOAA's head; basic
    cloud physics and dynamics be given higher priority;
    and that the legal, social, and economic impact of
    weather modification be thoroughly examined and
    appropriate regulatory and licensing legislation
    be sought.
(A Report to the President and the Congress, NACOA, June 29, 1973,
page viii)

We also believe it is particularly important that any such coordination
should be in the hands of a civilian agency; indeed, that all such research
should be conducted by civilian agencies.

Considerable doubt has been raised in the past over the nature of some
of the research conducted by the Defense Department in the area of
weather modification. You will recall the not too successful efforts
to increase rainfall over the Ho Chi Minh Trail several years ago at a
cost of $21.6 million. We have grave doubts about the merits of
any project such as this, but we are also concerned about the way in
which the incident was handled by the Government. The project was at
first flatly--and repeatedly--denied publicly and before Congress by
the Department of Defense, but the basic facts were ultimately conceded
some years later by former Defense Secretary Laird in a letter to the
Senate Foreign Relations Committee, which confirmed the allegations that
had been made.

Such incidents have given rise to continuing concern on our part over
the scope of federal research and development on environmental and
weather modification. What is significant about these incidents is that
they continue to occur in respect to Defense Department research, even
though DOD asserts such research has only peaceful applications,
such as airport fog dispersal. If this is the case, then it would
seem both logical and appropriate to place such research in civilian
agencies where it can be carried on with the same degree of precision
and success, since weapons' applications are not involved, and where it

The President
April 23, 1975
Page 4

would not cause new suspicions about the real nature of the work.

Weather modification is a field of great potential, promising considerable benefits to agriculture and transportation, to mention only two prime areas of research. At the same time the potential military applications of weather modification research are serious. Last summer's agreement with the Soviet Union to meet to discuss a ban on weather warfare is most encouraging. We hope that in the light of that agreement, you will be able to give favorable consideration to our recommendations.

Sincerely,

Gilbert Gude, M.C.   Claiborne Pell, U.S.S.   Donald M. Fraser, M.C.

(Courtesy of the office of U.S. Senator Claiborne Pell)

Appendix Six

METRIC SYSTEM

Unit	Abbreviation	Length Number of meters	Approximate U.S. equivalent
kilometer	km	1,000	0.62 mile
hectometer	hm	100	109.36 yards
decameter	dkm	10	32.81 feet
meter	m	1	39.37 inches

Unit	Abbreviation	Area Number of square miles	Approximate U.S. equivalent
square kilometer	sq km or km²	1,000,000	0.3861 square mile
hectare	ha	10,000	2.47 acres

Unit	Abbreviation	Mass and Weight Number of grams	Approximate U.S. equivalent
kilogram	kg	1,000	2.2046 pounds
hectogram	hg	100	3.527 ounces
decagram	dkg	10	0.353 ounce
gram	g or gm	1	0.035 ounce

Temperature Conversions

$$°Celsius = \frac{10(F-32)}{18} \qquad °Fahrenheit = \frac{18\,C}{10} + 32 \qquad Kelvin = C - 273$$

SELECTED BIBLIOGRAPHY

ABELSON, Philip H., "The World's Disparate Food Supplies," *Science*, January 24,1975.

ADELSON, Alan M., "Please Don't Steal the Atomic Bomb," *Esquire*, May 1969.

"Agricultural Productivity: On Borrowed Time?" *Science News*, January 18, 1975.

"Airplanes and Cancer: A New Jet Needed," *Science News*, April 5, 1975.

"Alaska Oil Field Supply Armada Blocked from Port as Openings in Ice Snap Shut," *Los Angeles Times*, September 1, 1975.

ALEXANDER, George, "The Aerosol Threat to Our Atmosphere," *Los Angeles Times*, April 27, 1975.

———, "Bible Flood Evidence Found," *Los Angeles Times*, September 25, 1975.

———, "Power Lines Cause 'Rain' of Electrons," *Los Angeles Times*, September 30, 1975.

———, "Scientist Links Ice Age to Sunspots," *Los Angeles Times*, August 24, 1975.

ALEXANDER, Tom, "Ominous Changes in the World's Weather," *Fortune*, February 1974; *Reader's Digest*, July 1974.

———, "What We Know—And Don't Know—About the Ozone Shield," *Fortune*, August 1975.

"Altering of Weather Widened Last Year to 5% of Nation," *New York Times*, July 13, 1975.

ANDERSON, Alan, Jr., "Forecast for Forecasting: Cloudy," *New York Times Magazine*, December 29, 1974.

ARDREY, Robert, *African Genesis*. New York: Athenum, 1961; Dell, 1967.

———, "The Glaciers Are Coming! . . ." *Playboy*, January 1976.

AREHART-TREICHEL, Joan, "Electromagnetic Pollution: Is It Hurting Our Health?" *Science News*, June 29, 1974.

———, "Green Revolution: Phase 2," *Science News*, July 21, 1973.

ASIMOV, Isaac, *The Ends of the Earth*. New York: Weybright & Talley, 1975.

———, "In The Game of Energy and Thermodynamics You Can't Even Break Even," *Smithsonian*, August 1970.

———, "Let's Suppose . . . A tale for the Year 3550 A.D.," *UNESCO Courier*, July–August 1974.

"Aswan's Impact," *Time*, May 5, 1975.

ATWATER, M. A., "Planetary Albedo Changes Due to Aerosols," *Science*, October 2, 1970.

AVSYUK, Grigori, and Vladimir KOTLYAKOV, "Glaciers on the Move," *UNESCO Courier*, June 1969.

AYERS, R. C., Jr., and S. MARTIN, et. al. "Oil Spills in the Arctic Ocean . . . Possibility of Large-Scale Thermal Effects," *Science*, November 29, 1974.

AYNSLEY, Eric, "How Air Pollution Alters Weather," in Arthur S. BOUGHEY, ed., *Readings in Man, the Environment, and Human Ecology*. New York: Macmillan, 1973.

BAHCALL, John N., "Neutrinos from the Sun," *Scientific American*, July 1969.

BAHM, Archie J., trans., *Tao Teh King by Lao Tzu*. New York; Ungar, 1958.

BAKER, D. James, Jr., "Models of Oceanic Circulation," *Scientific American*, January 1970.

BALZ, Dan, "Those Grain Deals: Are Controls Adequate?" *Los Angeles Times*, August 17, 1975.

BARBOUR, Ian G., ed., *Western Man and Environmental Ethics*. Reading, Mass.: Addison-Wesley, 1973.

BARDACH, John, *Harvest of the Sea*. New York: Harper & Row, 1968.

269

BARNEA, Joseph, "Geothermal Power," *Scientific American*, January 1972.

BARRY, Roger G., et al., "Continental Ice Sheets: Conditions for Growth," *Science*, December 5, 1975.

BASCOM, Willard, "The Disposal of Waste in the Ocean," *Scientific American*, August 1974.

BATTAN, Louis J., "Climate and Man," *Science*, October 24, 1969.

————, *Cloud Physics and Cloud Seeding*. New York: Anchor Books, 1962.

————, *Harvesting the Clouds*. New York: Doubleday, 1969.

————, *The Nature of Violent Storms*. New York: Doubleday, 1961.

————, "Some Problems in Changing the Weather" (mimeographed), American Meteorological Society, 1968.

————, *Weather*. Englewood Cliffs, N.J.: Prentice-Hall, 1974.

BAXTER, William J., *Warmer Weather!—Boom in North*. New York: International Economic Research Bureau, 1955.

BAZELL, Robert J., "Arid Land Agriculture . . . ," *Science*, March 12, 1971.

"Behind the Current Russian Grain Woes," *Time*, September 1, 1975.

BEHRMAN, Dan, "Making the Weather Pay Off," *UNESCO Courier*, August–September 1973.

————, *The New World of the Oceans*. Boston: Little, Brown, 1969.

BEKOFSKE, K., and V. C. LIU, "Internal Gravity Wave—Atmospheric Wind Interaction: A Cause of Clear Air Turbulence," *Science*, December 8, 1972.

BELL, Barbara, "Solar Variation as an Explanation of Climate Change," in Harlow SHAPLEY, ed., *Climatic Change*. Cambridge, Mass.: Harvard University Press, 1953.

BELLO, Francis, "Climate: The Heat May Be Off," *Fortune*, August 1954.

BERG, Alan, *The Nutrition Factor*. Washington, D.C.: Brookings Institution, 1973.

"Bering Straits 'Climate Factory' Urged by Russ," *Los Angeles Times*, February 13, 1956.

BERKNER, Lloyd V. and David BERGAMINI, "Climate to Suit Our Needs," *Science Digest*, November 1959.

BERRY, J. A., "Adaptation of Photosynthetic Processes to Stress," *Science*, May 9, 1975.

BHATNAGAR, P. L., "Internal Constitution of the Sun and Climatic Changes," in Harlow SHAPLEY, ed., *Climatic Change*. Cambridge, Mass.: Harvard University Press, 1953.

BICKEL, Lennard, *Facing Starvation: Norman Borlaug and the Fight Against Hunger*. New York: Reader's Digest Press, 1974.

"The Big Fallout from India's A-Test," *Newsweek*, June 3, 1974.

"Big Iran Uranium Purchase Told," *Los Angeles Times*, October 13, 1975.

BILLARD, Jules B., "Farming's Fantastic New Look," *National Geographic*, February 1970; *Reader's Digest*, June 1970.

BLOBAUM, Roger, "Corn Belt Study Puts Organic Farmers on Top!" *Organic Gardening and Farming*, October 1975.

————, "How China Uses Organic Farming Methods," *Organic Gardening and Farming*, July 1975.

BLUM, H. F., *Carcinogenesis by Ultraviolet Light*. Princeton, N.J.: Princeton University Press, 1959.

BLUMENSTOCK, David I., *The Ocean of Air*. New Brunswick, N.J.: Rutgers University Press, 1959.

" 'Body Coolant' Called Key to 200-Year Life," *Los Angeles Times*, March 24, 1974.

BOERMA, Addeke H., "A World Agricultural Plan," *Scientific American*, August 1970.

BOFFEY, Philip M., "Sea-Level Canal . . . ," *Science*, January 29, 1971.
BORGESE, Elisabeth Mann, ed., *Pacem in Maribus*. New York: Dodd, Mead, 1972.
BORGSTROM, Georg, *The Food and People Dilemma*. North Scituate, Mass.: Duxbury
    Press, 1973.
———, *Harvesting the Earth*. New York: Abelard-Schuman, 1973.
———, *The Hungry Planet*, 2nd rev. ed. New York: Macmillan, 1972.
———, *Too Many*. New York: Macmillan, 1969.
BOUGHEY, Arthur S., ed., *Readings in Man, The Environment, and Human Ecology*.
    New York: Macmillan, 1973.
BOURGHOLTZER, Frank, "Glacier Poses Threat to Valdez' Ice-Free Port," *Los
    Angeles Times*, September 22, 1975.
BOVA, Ben, *The Fourth State of Matter*. New York: St. Martin's Press, 1971; Mentor,
    1974.
BRAY, J. Roger, "Solar-Climate Relationships in the Post-Pleistocene," *Science*, March
    26, 1971.
BREYER, Siegfried, *Guide to the Soviet Navy*. Annapolis, Md.: U.S. Naval Institute,
    1970.
BRIER, Glenn W., "Design and Evaluation of Weather Modification Experiments," in
    W. N. Hess, ed., *Weather and Climate Modification*. New York: Wiley, 1974.
BROECKER, Wallace S., "Absolute Dating and the Astronomical Theory of Glacia-
    tion," *Science*, January 21, 1966.
———, "Climatic Change: Are We on the Brink of a Pronounced Global Warming?"
    Science, August 8, 1975.
———, "Floating Glacial Ice Caps in the Arctic Ocean," *Science*, June 13, 1975.
BROOKS, C. E. P., *Climate in Everyday Life*. New York: Philosophical Library, 1951.
———, *Climate Through The Ages*, 2nd rev. ed. New York: Dover, 1970.
BROOKS, Paul, "The Plot to Drown Alaska," *The Atlantic*, May 1965.
BROUWER, Dirk, "The Polar Motion and Changes in the Earth's Orbit," in Harlow
    SHAPLEY, ed., *Climatic Change*. Cambridge, Mass.: Harvard University Press,
    1953.
BROWN, Lester R., *In The Human Interest*. New York: Norton, 1974.
———, *Seeds of Change*. New York: Praeger, 1970.
———, "Who Will Get the One Extra Crust of Bread?" *Los Angeles Times*, November
    5, 1974.
———, "The World Food Prospect," *Science*, December 12, 1975.
BROWN, Lester R., and Erik P. ECKHOLM, *By Bread Alone*. New York: Praeger,
    1974.
BROWN, Lester R., and Gail W. FINSTERBUSH, *Man and His Environment: Food*.
    New York: Harper & Row, 1972.
BROWN, N. L., and E. R. PARISER, "Food Science in Developing Countries,"
    Science, May 9, 1975.
BRYSON, Reid A., "All Other Factors Being Constant . . ." *Weatherwise*, April 1968.
———, *Climatic Modification by Air Pollution, II: The Sahelian Effect*. Madison, Wis.:
    Institute for Environmental Studies, 1973.
———, *Heyuppskera: An Heuristic Model For Hay Yield in Iceland*. Madison, Wis.:
    Institute for Environmental Studies, 1974.
———, "Is Man Changing the Climate of Earth?" *Saturday Review*, April 1, 1967.
———, *The Lessons of Climatic History*. Madison, Wis.: Institute for Environmental
    Studies, 1974.
———, "A Perspective on Climatic Change," *Science*, May 17, 1974.
BRYSON, Reid, and David BAERREIS, "Possibilities of Major Climatic Modification

and Their Implications," *Bulletin of the American Meteorological Society*, March 1967.

BUCHAN, Alistair, ed., *A World of Nuclear Powers?* Englewood Cliffs, N.J.: Prentice-Hall, 1966.

BUDYKO, Mikhail, *Climate and Life*. Leningrad: Hydrological Publishing House, 1971; New York: Academic Press, 1974.

———, "The Effect of Solar Radiation Variations on the Climate of the Earth," *Tellus*, Vol. 21.

———, *The Heat Budget of the Earth*. Leningrad: Hydrological Publishing House, 1963.

———, "Will Human Activity Overheat the Earth?" *Christian Science Monitor*, August 27, 1974; *Current*, October 1974.

BUTZER, K. W., "Climatic Change," in *Encyclopedia Britannica*, Macropedia, Vol. 4, 1974.

BYLINSKY, Gene, "A New Scientific Effort to Boost Food Output," *Fortune*, June 1975.

CAIDIN, Martin, *Hydrospace*. New York: Dutton, 1964.

CALDER, Nigel, *Eden Was No Garden*. New York: Viking, 1967.

———, ed., *Nature In The Round*. New York: Viking, 1973.

———, *The Restless Earth*. New York: Viking, 1972.

———, ed., *Unless Peace Comes*. New York: Viking, 1968.

———, *The Weather Machine*. New York: Viking, 1975.

CALVERT, James F., "Up Through the Ice of the North Pole," *National Geographic*, July 1959.

CAMPBELL, W. J., and S. MARTIN, "Oil and Ice in the Arctic Ocean: Possible Larger-Scale Interactions," *Science*, July 6, 1973.

CAMERON, I. R., "Meteorites and Cosmic Radiation," *Scientific American*, July 1973.

CANBY, Thomas Y., "Can The World Feed Its People?" *National Geographic*, July 1975.

"Carbon Dioxide: Lost and Found," *Science News*, April 19, 1975.

CARLSON, P. S., and J. C. POLACCO, "Plant Cell Cultures: Genetic Aspects of Crop Improvement," *Science*, May 9, 1975.

CARTER, Luther J., "Environmental Law . . . ," *Science*, March 23 & 30, 1973.

———, "Federal Control of Rainmakers . . . ," *Science*, June 29, 1973.

———, "Icebergs and Oil Tankers: USGS Glaciologists are Concerned," *Science*, November 14, 1975.

———, "Weather Modification: Colorado Heeds Voters in Valley Dispute," *Science*, June 29, 1973; reader response, August 31, 1973.

CARUTHERS, Osgood, "Russian Bids U.S. Help in Bering Dam," *New York Times*, October 25, 1959.

CHANDLER, T. J., *The Air Around Us*. New York: Natural History Press, 1969.

CHANG, Jen-hu, *Climate and Agriculture*. Chicago: Aldine, 1968.

"Change Tropical Weather by Salt-Seeding Trade Winds," *Science News Letter*, November 3. 1951.

CHAPIN, Henry, and F. G. Walton SMITH, *The Ocean River*. New York: Scribner's, 1962.

CHARLSON, R. J., et al., "Aerosol Concentrations: Effect on Planetary Temperatures," *Science*, January 7, 1972; response by S. I. RASOOL and S. H. SCHNEIDER, same issue.

CHARNEY, Jule, et al., "Drought in the Sahara: A Biogeophysical Feedback Mechanism," *Science*, February 7, 1975.

CHESTERMAN, John, et al., *An Index of Possibilities: Energy and Power*. New York: Pantheon, 1974.

CHYLEK, P., and J. A. COAKLEY, Jr., "Aerosols and Climate," *Science*, January 11, 1974.

CLAIBORNE, Robert, *Climate, Man, and History*. New York: Norton, 1970.

CLARK, Colin, *Starvation or Plenty?* New York: Taplinger, 1970.

CLARK, John R., "Thermal Pollution and Aquatic Life," *Scientific American*, March 1969.

CLARK, Wilson, *Energy For Survival*. New York: Anchor Press, 1974.

CLARKE, Arthur C., *The Challenge of the Sea*. New York: Holt, Rinehart and Winston, 1960; Dell, 1966.

"Climate Vs. Magnetism: Short-term Match," *Science News*, March 17, 1973.

Climatic Change. World Meteorological Organization, 1971.

"Climatic Effects of Particulates," *Science News*, March 14, 1970.

COCHRANE, Willard W., *The World Food Problem*. New York: Crowell, 1969.

COHEN, Daniel, "Should We Change the Weather?" *Science Digest*, November 1962.

COHN, Victor, "Weather War: A Gathering Storm," *Washington Post*, July 2, 1972.

"Cold Forecast for Next 50 Years," *Electronics & Power*, July 12, 1973.

COLE, H. S. D., et al., eds., *Models of Doom*. New York: Universe Books, 1973.

COLLIGAN, Douglas, "Brace Yourself for Another Ice Age," *Science Digest*, February 1973.

"Colonizing Space," *Time*, May 26, 1975.

COMMONER, Barry, *The Closing Circle*. New York: Knopf, 1971.

———, *Science and Survival*. New York: Viking, 1966.

CONGER, Dean, "Siberia: Russia's Frozen Frontier," *National Geographic*, March 1967.

CONINE, Ernest, "High Oil Prices Will Breed Nuclear Arms," *Los Angeles Times*, August 22, 1975.

CONRAD, Victor, "Climatic Changes or Cycles?" in Harlow SHAPLEY, ed., *Climatic Change*. Cambridge, Mass.: Harvard University Press, 1953.

CONWAY, John A., "Nuclear Bookkeeping," *Newsweek*, August 18, 1975.

———, "Nukes on the Nile," *Newsweek*, August 25, 1975.

COOK, Don, "Debate Over New Ice Age Heating Up," *Los Angeles Times*, September 7, 1975.

"Cooler or Warmer?" *Science News*, July 8, 1972.

COON, Carleton S., "Climate and Race," in Harlow SHAPLEY, ed., *Climatic Change*. Cambridge, Mass.: Harvard University Press, 1953.

COPELAND, Miles, *Without Cloak or Dagger*. New York: Simon and Schuster, 1974.

COURT, Arnold, ed., *Eclectic Climatology*. Corvallis, Ore.: Oregon State University Press, 1968.

COWAN, Clyde L., *Ocean Currents and Climate*. (Monograph). Washington, D.C., 1965.

COWEN, Robert C., *Frontiers of the Sea*, rev. ed. New York: Doubleday, 1969.

———, "More Variable Weather Threatens World Food Supply," *Christian Science Monitor*, December 9, 1975.

COX, Allan, et al., "Reversals of the Earth's Magnetic Field," *Scientific American*, February 1967.

COX, Jeff, "Factory Farming is Not Efficient," *Organic Gardening*, June 1973.

CRAIG, Richard A., *The Edge of Space*. New York: Doubleday, 1968.

CRANDALL, Willard A., "Electric Power General: Options for Environmental Improvement," in William H. MATTHEWS, et al., eds., *Man's Impact on Climate*. Cambridge, Mass.: MIT Press, 1971.

CRITCHFIELD, Howard J., *General Climatology*. Englewood Cliffs, N.J.: Prentice-Hall, 1966.

CRONIN, John F., "Recent Volcanism and the Stratosphere," *Science*, May 21, 1971.

CROSSON, P. R., "Institutional Obstacles to Expansion of World Food Production," *Science*, May 9, 1975.

CRUTCHFIELD, James A., "Economic Evaluation of Weather Modification," in Robert G. FLEAGLE, ed., *Weather Modification*. Seattle: University of Washington Press, 1969.

"Currents in Arctic Believed to Play Key Role in Climate," *New York Times*, March 13, 1975.

"Damming the Icy Seas," *Newsweek*, September 29, 1958.

DANSGAARD, W., et al., "Climatic Record Revealed by the Camp Century Ice Core," in Karl K. TUREKIAN, ed., *Late Cenozoic Glacial Ages*. New Haven: Yale University Press, 1971.

———, "One Thousand Years of Climatic Record from Camp Century on the Greenland Ice Sheet," *Science*, October 17, 1969.

DANSGAARD, W., and Henrik TAUBER, "Glacier Oxygen-18 Content and Pleistocene Ocean Temperatures," *Science*, October 24, 1969.

DARDEN, Lloyd, *The Earth in the Looking Glass*. New York: Anchor Press, 1974.

DARLING, Fraser, and John P. MILTON, eds., *Future Environments of North America*. New York: Natural History Press, 1966.

DARMSTADTER, Joel, et al., *Energy in the World Economy*. Baltimore: Johns Hopkins, 1971.

DAS, P. K., *The Monsoons*. London: Arnold, 1972.

DAVIS, Kenneth S., and John Arthur DAY, *Water: The Mirror of Science*. New York: Anchor Books, 1961.

DAVIS, Kingsley, "The Migrations of Human Population," *Scientific American*, September 1974.

DAVIS, Ray J., "Weather Modification Litigation and Statutes," in W. N. HESS, ed., *Weather and Climate Modification*. New York: Wiley, 1974.

DAY, John A., *The Science of Weather*. Reading, Mass.: Addison-Wesley, 1966.

DAY, John A., and Gilbert L. STERNES, *Climate and Weather*. Reading, Mass.: Addison-Wesley, 1970.

DAY, Samuel H., Jr., "Prospects of Nuclear Violence: The Drift Becomes a Rush," *Bulletin of Atomic Scientists*, September 1975.

"Death as the Compass Swings," *New Scientist*, June 8, 1967.

De BELL, Garrett, ed., *The Environmental Handbook*. New York: Ballantine, 1970.

DEMENY, Paul, "The Populations of Underdeveloped Countries," *Scientific American*, September 1974.

DENTON, George H., and Stephen C. PORTER, "Neoglaciation," *Scientific American*, June 1970.

"Development May Not Cut Population," *Science News*, February 1, 1975.

DICK, William, and Granville TOOGOOD, "Harvard Professor Warns of . . . Doomsday Weapon," *National Enquirer*, June 17, 1975.

DINGLE, A. Nelson, "Man-Made Climatic Changes: Seeding by Contrails," *Science*, July 30, 1971.

"Dirty Grain," *Time*, June 30, 1975.

D'MONTE, Darryl, and K. R. Sundar RAJAN, "India's Nuclear and Political Dilemmas," *Los Angeles Times*, June 6, 1975.

DOBSON, G. M. B., *Exploring the Atmosphere*, 2nd ed. London: Oxford University Press, 1968.

DONN, William L., and Maurice EWING, "A Theory of Ice Ages, III," *Science*, June 24, 1966.

DONOVAN, Robert J., "World Security Depends More on Food than Armies," *Los Angeles Times*, December 19, 1974.

DOUGLAS, John H., "Climate Change: Chilling Possibilities," *Science News*, March 1, 1975.

_____, "Confronting Famine," *Science News*, May 18, 1974.

_____, "The Great Grain Game," *Science News*, November 30, 1974.

_____, "The Invisible Famine," *Science News*, December 13, 1975.

_____, "The New Green Revolution," *Science News*, October 5, 1974.

_____, "The Omens of Famine," *Science News*, May 11, 1974.

DRESCH, Jean, "Drought Over Africa," *UNESCO Courier*, August–September 1973.

DROSNIN, Michael, "Not with a Bang, but with a Pssssst!" *New Times*, March 7, 1975.

"Drought: A Southward Shift of World Climate," *Science News*, September 29, 1973.

DRYSSEN, D., and D. JAGNER, eds., *The Changing Chemistry of the Oceans*. New York: Wiley, 1972.

DUMOND, D. E., "The Limitation of Human Population . . . ," *Science*, February 28, 1975.

DUMONT, René, "A World Gone Mad," *UNESCO Courier*, July–August 1974.

DUMONT, René, and Bernard ROSIER, *The Hungry Future*. New York: Praeger, 1969.

DUPLESSY, Jean Claude, et al., "Weyl's Theory of Glaciation Supported by Isotopic Study of Norwegian Core K 11," *Science*, June 20, 1975.

DWYER, J. T., and J. MAYER, "Beyond Economics and Nutrition: The Complex Basis of Food Policy," *Science*, May 9, 1975.

DYSON, James L., *The World of Ice*. New York: Knopf, 1962.

"The Earth Glitch That Wasn't," *Science News*, September 1, 1973.

"Earth's Magnetic Field Has Its Own Axis," *Science News*, September 22, 1973.

"Earth's Plantlife: Ultraviolet Peril," *Science News*, August 23 & 30, 1975.

EBERHART, Jonathan, "And Into the Warming Sea Rode the 4,000," *Science News*, June 1, 1974.

_____, "Home From the Warming Sea—And Into the Computer," *Science News*, November 23, 1974.

_____, "A Magnet Named Earth," *Science News*, May 24, 1975.

EDMONDSON, W. T., "Ecology and Weather Modification," in Robert G. FLEAGLE, ed., *Weather Modification*. Seattle: University of Washington Press, 1969.

"Effects of Pollution," *Science News*, April 24, 1971.

EHRLICH, Paul R., *The End of Affluence*. New York: Ballantine, 1974.

_____, *The Population Bomb*. New York: Ballantine, 1968.

EHRLICH, Paul R., and Anne H., EHRLICH, *Population, Resources, Environment*. San Francisco: Freeman, 1972.

EHRLICH, Paul R., and Richard L. HARRIMAN, *How to Be A Survivor*. New York: Ballantine, 1971.

EISELEY, Loren, *The Unexpected Universe*. New York: Harcourt, Brace & World, 1969.

ELEGANT, Robert S., "Spur to Nuclear Race Seen in Viet Fall," *Los Angeles Times*, May 25, 1975.

ELLIOTT, Laurence H., *Climate and Man*. New York: McGraw-Hill, 1969.

EMERY, K. O., "Sea Levels 7,000 to 20,000 Years Ago," *Science*, August 11, 1967.

EMILIANI, Cesare, "Ancient Temperatures," *Scientific American*, February 1958.

_____, "Isotopic Paleotemperatures," *Science*, November 18, 1966; debated, with rebuttal by Emiliani, *Science*, August 11, 1967.

_____, "Pleistocene Temperatures," *Journal of Geology*, November 1955.

_____, "Quaternary Paleotemperatures and the Duration of the High Temperature Intervals," *Science*, October 27, 1972.

EMILIANI, Cesare, et al., "Paleoclimatological Analysis of Late Quaternary Cores from the Northeastern Gulf of Mexico," *Science*, September 26, 1975.

"An End to the 'Little Ice Age'?" *Science News*, August 16, 1975.

"Energy Goal: 25 percent Solar by 2020," *Science News*, August 23 & 30, 1975.

"Energy: Inflation for Some, Disaster for Others," *Science News*, October 5, 1974.

ENGLER, Robert, *The Politics of Oil*. Chicago: University of Chicago Press, 1967.

ENTERLINE, James Robert, *Viking America*. New York: Doubleday, 1972.

EPSTEIN, William, "The Proliferation of Nuclear Weapons," *Scientific American*, April 1975.

ERICKSON, E. W., and L. WAVERMAN, eds., *The Energy Question*. Toronto: University of Toronto Press, 1974.

ERICSON, David B., and Goesta WOLLIN, *The Deep and the Past*, New York: Knopf, 1964.

_____, "Pleistocene Climates and Chronology in Deep-Sea Sediments," *Science*, December 13, 1968.

"Eruptions and Climate: Proof from an Icy Pudding," *New Scientist*, February 17, 1972.

"Evidence for Weakening Gravity," *Science News*, April 13, 1974.

EWING, Maurice, and William L. DONN, "A Theory of Ice Ages," *Science*, June 15, 1956.

_____, "A Theory of Ice Ages, II," *Science*, May 16, 1958.

"Excerpts From Secretary Kissinger's Address to the Special U.N. Assembly," *New York Times*, April 16, 1974.

EZRA, A. A., "Technology Utilization: Incentives and Solar Energy," *Science*, February 28, 1975.

FAIRBRIDGE, Rhodes W., "The Changing Level of the Sea," *Scientific American*, May 1960.

_____, "Climatology of a Glacial Cycle," *Quarternary Research*, November 1972.

_____, ed., *The Encyclopedia of Oceanography*. New York: Reinhold, 1966.

_____, ed., *Solar Variations, Climatic Change and Related Geophysical Problems*. New York Academy of Sciences, 1961.

FAIRHALL, David, *Russian Sea Power*. Boston: Gambit, 1971.

FAIRLIE, Henry, "The Fallacy of 'Propitious Circumstances,' " *Harper's Magazine*, January 1975.

FALK, Richard A., *Environmental Warfare and Ecocide—Facts, Appraisal, and Proposals*, rev. ed., in *Prohibiting Military Weather Modification*, hearings before the Oceans and International Environment Subcommittee of the U.S. Senate Committee on Foreign Relations, 92nd Congress, Second Session, July 26 and 27, 1972.

_____, *This Endangered Planet*. New York: Random House, 1971.

"Famine Fears Rise, Battle Lines Form," *Science News*, July 27, 1974.

FARRINGTON, Daniel, Jr., et al., "Sunburn," *Scientific American*, July 1968.

FARVAR, M. Taghi, and John P. MILTON, eds., *The Careless Technology: Ecology and International Development*. New York: Natural History Press, 1972.

"Federal Control of the Rainmakers?" *Science*, June 29, 1973.

FEDOROV, Yevgeny K., "Modification of Meteorological Processes," in W. N. HESS, ed., *Weather and Climate Modification*. New York: Wiley, 1974.

"Feeding a Hungry World . . ." (interview with Norman Borlaug and Georg Borgstrom), *Los Angeles Times*, April 23, 1972.

FEININGER, Tomás, "Less Rain in Latin America" (letter), *Science*, April 5, 1968.

Bibliography   277

FELD, Bernard T., "Making the World Safe for Plutonium," *Bulletin of Atomic Scientists*, May 1975.

FISHER, Arthur, "Energy From the Sea . . ." (series) *Popular Science*, May, June, July, 1975.

FISHER, C. R., et al., "EMP and the Radio Amateur," *QST*, September 1975.

FLEAGLE, Robert G., "BOMEX: An Appraisal of Results," *Science* June 9, 1972.

———, *Weather Modification*. Seattle: University of Washington Press, 1969.

FLEAGLE, Robert G., and Joost A. BUSINGER, "The 'Greenhouse Effect,' " (letter), *Science*, December 12, 1975.

FLETCHER, Joseph O., "Controlling the Planet's Climate," *Impact of Science on Society*, Vol. 19.

———, *The Heat Budget of the Arctic Basin and Its Relation to Climate*. Santa Monica, Cal.: RAND Corporation, 1965.

———, "Polar Ice and the Global Climate Machine," *Bulletin of Atomic Scientists*, December 1970.

FLÖHN, Hermann, *Climate and Weather*. New York: McGraw-Hill, 1969.

"Fluorocarbons and Ozone . . . ," *Science News*, October 5, 1975.

"Food" *Mosaic* (National Science Foundation), Special Issue, May–June 1975.

"Food: 'A Good Moment for Stocktaking,' " *Science News*, May 10, 1975.

"Food and Population: Thinking the Unthinkable," *Science News*, November 30, 1974.

"Food: The Assessment," *Science News*, November 16, 1974.

"Food Conference: Let them eat Words?" *Science News*, November 2, 1974.

"Food Conference: New Beginnings," *Science News*, November 23, 1974.

"Food, Famine and the Rome Conference," *Science News*, November 9, 1974.

"Food Weapon Unsheathed," *Science News*, December 14, 1974.

FOREIGN POLICY ASSOCIATION, *Toward the Year 2018*. New York: Cowles Education Corporation, 1968.

FORRESTER, Jay W., "The Computer and Social Catastrophe," *Intellectual Digest*, November 1971.

———, *World Dynamics*. Cambridge, Mass.: Wright-Allen, 1971.

FOWLER, William A., "The Case of the Missing Neutrinos," in *Science Year 1974*. Chicago: Field Enterprises, 1973.

———, "What Cooks with Solar Neutrinos?" *Nature*, July 7, 1972.

FOWLER, W. B., and J. D. HELVEY, "Effect of Large-Scale Irrigation on Climate in the Columbia Basin," *Science*, April 12, 1974.

FRANKEL, O. H., et al., "Genetic Dangers in the Green Revolution," *Ceres*, September–October 1969.

FRANKENBERG, Theodore T., "Air Pollution from Power Plants and Its Control," in Arthur S. BOUGHEY, ed., *Readings in Man, the Environment, and Human Ecology*. New York: Macmillian, 1973.

FRANZ, Maurice, "1976—It's Going to Get Colder," *Organic Gardening and Farming*, January 1976.

FRAZIER, Kendrick, "The Specter of Meteorological Warfare," *Science News*, July 15, 1972.

FREEMAN, Orville, "Malthus, Marx and the North American Breadbasket," *Foreign Affairs*, July 1967.

FREEMAN, Ronald, and Bernard BERELSON, "The Human Population," *Scientific American*, September 1974.

FREEMAN, S. David, *Energy: The New Era*. New York: Vintage, 1974.

"Freon: Destroying the Ozone Layer?" *Science News*, September 21, 1974.

FRIEDAN, Betty, "The Coming Ice Age," *Harper's Magazine*, September 1958.

FRIEDMAN, Jaime, "Rationing Warmth," *National Observer*, October 11, 1975.

FRIEDMANN, Wolfgang, *The Future of the Oceans*. New York: Braziller, 1971.

FRITTS, Harold C., "Growth Rings of Trees: Their Correlation with Climate," *Science*, November 25, 1966.

_____, "Tree Rings and Climate," *Scientific American*, May 1972.

"From Forest to Fuel: The Mounting Crisis," *Science News*, September 27, 1975.

FULLER, R. Buckminister, *Ideas and Integrities*. Englewood Cliffs, N.J.: Prentice-Hall, 1963; Collier Books, 1969.

_____, *Planetary Planning*. New Delhi: Indraprastha Press, 1969.

FULLER, R. Buckminster, Jerome AGEL, and Quentin FIORE, *I Seem To Be A Verb*. New York: Bantam, 1970.

FURNEAUX, Rupert, *Krakatoa*. Englewood Cliffs, N.J.: Prentice-Hall, 1964.

GAGIN, A., and J. NEUMANN, "Rain Stimulation and Cloud Physics in Israel," in W. N. HESS, ed., *Weather and Climate Modification*. New York: Wiley, 1974.

GARDNER, Hugh, "The Stormy Saga of Weather Mod," *New Times*, June 27, 1975.

GARLICK, J. P., and A. W. J. KEATS, eds., *Human Ecology in the Tropics*. Oxford: Pergamon Press, 1970.

GASKELL, T. F., *The Gulf Stream*. New York: John Day, 1973; Mentor, 1974.

_____, *Physics of the Earth*. New York: Funk & Wagnalls, 1970.

GATES, David M., *Man and His Environment: Climate*. New York: Harper & Row, 1972.

GATES, W. Lawrence, "Understanding Climatic Change," in *RAND 25th Anniversary Volume*. Santa Monica, Cal.: RAND Corporation, 1973.

GAVAN, James D., and John DIXON, "India: A Perspective on the Food Situation," *Science*, May 9, 1975.

"Geomagnetic Storms May Influence Atmosphere," *Science News*, March 17, 1973.

GEYELIN, Philip, *Lyndon B. Johnson and the World*. New York: Praeger, 1966.

GILFILLAN, Edward, *Migration to the Stars*. New York: Luce, 1975.

GILLETTE, Robert, "India: Into the Nuclear Club on Canada's Shoulders," *Science*, June 7, 1974.

_____, "Uranium Enrichment: Rumors of Israeli Progress . . . ," *Science*, March 22, 1974.

_____, "Uranium Enrichment: With Help, South Africa Is Progressing," *Science*, June 13, 1975.

GILLILAND, Martha W., "Energy Analysis and Public Policy," *Science*, September 26, 1975.

GINZBURG, V. L., "The Astrophysics of Cosmic Rays," *Scientific American*, February 1969.

GLASS, Billy, and Bruce C. HEEZEN, "Tektites and Geomagnetic Reversals," *Nature*, April 22, 1967; *Scientific American*, July 1967.

GLUECK, Nelson, *Rivers in the Desert*. New York: Farrar, Straus & Cudahy, 1959.

GOLDMAN, Marshall I., ed., *Ecology and Economics*. Englewood Cliffs, N.J.: Prentice-Hall, 1972.

GOODY, Richard M., and James C. G. WALKER, *Atmospheres*. Englewood Cliffs, N.J.: Prentice-Hall, 1972.

GORDON, Arnold L., and Hoyt W. TAYLOR, "Seasonal Change of Antarctic Sea Ice Cover," *Science*, January 31, 1975.

GORDON, Theodore J., "Some Crises That Will Determine the World of 1994," *The Futurist*, June 1974; *Current*, December 1974.

GRANT, James P., "Poor Nations Have Legitimate Grievances," *Los Angeles Times*, October 5, 1975.

GRAVES, E., "Nuclear Excavation of a Sea-Level, Isthmian Canal," in *Engineering With Nuclear Explosives.* Washington, D.C.: Atomic Energy Commission, 1964.

GRECO, R. V., and R. E. TURNER, "Cloud Physics Laboratory: A Step Toward Weather Control," *Astronautics & Aeronautics,* March 1975.

GREGG, Michael C., "The Microstructure of the Ocean," *Scientific American,* February 1973.

GREENFIELD, Jerome, *Wilhelm Reich Vs. The U.S.A.* New York: Norton, 1974.

GREENFIELD, S. M., *Weather Modification Research: A Desire and an Approach.* Santa Monica, Cal.: RAND Corporation, 1969.

GREENFIELD, S. M., et al., *A New Rational Approach to Weather Control Research.* Santa Monica, Cal.: RAND Corporation, 1962.

GRIBBIN, John R., *Forecasts, Famines, and Freezes.* London: Wildwood House, 1975; New York: Walker, 1976 (Tentative).

————, "Our Weather: A Link With the Planets?" *New Scientist,* December 27, 1973.

————, "Whither the Weather?" *Natural History,* June–July 1975.

GRIBBIN, John R., and Stephen H. PLAGEMANN, *The Jupiter Effect.* New York: Walker, 1974; Vintage, 1975.

GULLION, Edmund A., ed., *Uses of The Seas.* Englewood Cliffs, N.J.: Prentice-Hall, 1968.

GWERTZMAN, Bernard, "A U.S.–Soviet Ban on Weather Use for War Is Near," *New York Times,* June 24, 1975.

GWYNNE, Peter, "The Cooling World," *Newsweek,* April 28, 1975.

HAAG, William G., "The Bering Strait Land Bridge," *Scientific American,* January 1962.

HABER, Heinz, *Our Blue Planet.* New York: Scribner's, 1969.

HACKER, David W., "A Lawsuit Grows from Cloud Seeds," *National Observer,* June 7, 1975.

HADLOW, Leonard, *Climate, Vegetation, and Man.* Westport, Conn.: Greenwood Press, 1952.

HAERENDEL, Gerhard, and Reimar LÜST, "Artificial Plasma Clouds in Space," *Scientific American,* November 1968.

HALACY, D. S., Jr., *The Coming Age of Solar Energy,* rev. ed. New York: Harper & Row, 1973.

————, *The Geometry of Hunger.* New York: Harper & Row, 1972.

————, *The Weather Changers.* New York: Harper & Row, 1968.

HALSTED, Thomas A., "The Spread of Nuclear Weapons—Is the Dam About to Burst?" *Bulletin of Atomic Scientists,* May 1975.

HAMER, John, *World Weather Trends.* Washington, D.C.: Editorial Research Reports, 1974.

HAMILTON, Roger, "Can We Harness the Wind?" *National Geographic,* December 1975.

HAMMOND, Allen L., "Crop Forecasting from Space," *Science,* May 2, 1975.

————, "Global Meteorology . . . ," *Science,* October 15 & 22, 1971.

————, "Global Weather Experiment: The Petrodollar Connection," *Science,* June 20, 1975.

————, "Hurricane Prediction and Control . . . ," *Science,* August 17, 1973.

————, "Long-Range Weather Forecasting; Sea Temperature Anomalies," *Science,* June 7, 1974.

————, "Modeling the Climate: A New Sense of Urgency," *Science,* September 27, 1974.

————, "NOAA to Try Rescue Rainmaking," *Science,* April 2, 1971.

_____, "Probing the Tropical Firebox: International Atmospheric Science," *Science*, June 20, 1975.

_____, "Solar Energy: The Largest Resource," *Science*, September 22, 1972.

_____, "Solar Neutrinos: Where Are They?" *Science*, February 4, 1972.

_____, "The Uniqueness of the Earth's Climate," *Science*, January 24, 1975.

_____, "Weather Modification—A Technology Coming of Age," *Science*, May 7, 1971.

HAMMOND, Allen L., and Thomas H. MAUGH, III, "Stratospheric Pollution: Multiple Threats to Earth's Ozone," *Science*, October 25, 1974.

HAMMOND, George S., and W. Murray TODD, "Technical Assistance and Foreign Policy," *Science*, September 26, 1975.

HANSON, William B., "Earth's Dynamic Thermosphere, " *Astronautics & Aeronautics*, January 1975.

HAPGOOD, Charles H., *The Path of the Pole*. Philadelphia: Chilton, 1970.

HARDIN, Clifford M., ed., *Overcoming World Hunger*. Englewood Cliffs, N.J.: Prentice-Hall, 1969.

HARDIN, Garrett, *Exploring New Ethics for Survival: The Voyage of the Spaceship Beagle*. New York: Viking, 1972.

_____, "Nobody Ever Dies of Overpopulation," *Science*, February 12, 1971.

_____, ed., *Population, Evolution, and Birth Control*, 2nd ed. San Francisco: Freeman, 1969.

HARDY, R. W. F., and U. D. HAVELKA, "Nitrogen Fixation Research: A Key to World Food?" *Science*, May 9, 1975.

HARE, F. K., *The Restless Atmosphere*, 3rd ed. New York: Harper & Row, 1963.

HARKABI, Y., *Nuclear War and Nuclear Peace*. Jerusalem: Israel Program for Scientific Translations, 1966.

HARLAN, J. R., "Our Vanishing Genetic Resources," *Science*, May 9, 1975.

HARRAR, J. George, *Strategy Toward the Conquest of Hunger*. New York: Rockefeller Foundation, 1967.

HARRIS, David R., "New Light on Plant Domestication and the Origins of Agriculture," *Geographical Review*, January 1967.

HARTMANN, William K., "If Mars Once Had Rivers, Was the Sun Then Hotter?" *Smithsonian*, September 1975.

HAURWITZ, F., et al., "Distribution of Tropospheric Planetary Radiation in the Southern Hemisphere," *Journal of Applied Meteorology*, June 1974.

HAUSER, Philip M., ed., *The Population Dilemma*, 2nd ed. Englewood Cliffs, N.J.: Prentice-Hall, 1963.

HAYS, James D., "The Ice Age Cometh," *Saturday Review*, April 1973.

HEEZEN, Bruce C., and Ian D. MacGREGOR, "Riddles Chalked on the Ocean Floor," *Saturday Review*, February 19, 1972.

HEIDEL, Karen, "Turbidity Trends at Tucson, Arizona," *Science*, September 1972.

HEILBRONER, Robert L., *An Inquiry Into the Human Prospect*. New York: Norton, 1974.

HELLMAN, Hal, *Energy in the World of the Future*. New York: Evans, 1973.

HENDRICKS, Sterling B., "How Light Interacts With Living Matter," *Scientific American*, September 1968.

HERRING, Peter J., and Malcolm R. CLARKE, eds., *Deep Oceans*. New York: Praeger, 1971.

HERSH, Seymour M., *Chemical and Biological Warfare*. Indianapolis: Bobbs-Merrill, 1968.

_____, "Rainmaking Is Used as Weapon by U.S. . . . ," *New York Times*, July 3, 1972.

HESS, Wilmot N., "Man-Made Radiation Belts," *International Science and Technology*, September 1963.

————, ed., *Weather and Climate Modification*. New York: Wiley, 1974.

HILL, Gladwin, "A Sterility Drug in Food is Hinted," *New York Times*, November 25, 1969.

HOAD, Brian, "The Climate Is/Is Not Changing," *Atlas World Press Review*, August 1974.

HOFFERT, Martin I., and Richard W. STEWART, "Stratospheric Ozone—Fragile Shield?" *Astronautics & Aeronautics*, October 1975.

HOLDREN, John P., and Paul R. EHRLICH, "Human Population and the Global Environment," *American Scientist*, May–June 1974.

————, "Man Is Courting Ecological Disaster," *UNESCO Courier*, July–August 1974.

HOLDREN, John, and Philip HERRERA, *Energy*. San Francisco: Sierra Club, 1971.

HOLT, S. J., "The Food Resources of the Ocean," *Scientific American*. September 1969.

"Honduras: Rustlers in the Sky," *Time*, August 15, 1949.

HOULT, David P., ed., *Oil On The Sea*. New York: Plenum, 1969.

"How The Ice Age Began," *Time*, March 24, 1975.

HOWARD, R. A., et al., "The Decision to Seed Hurricanes," *Science*, June 16, 1972.

HSÜ, Kenneth J., "When the Mediterranean Dried Up," *Scientific American*, December 1972.

HULSE, Joseph H., and David SPURGEON, "Triticale," *Scientific American*, August 1974.

HUNT, Robert S., "Weather Modification and the Law," in Robert G. FLEAGLE, ed., *Weather Modification*. Seattle: University of Washington Press, 1969.

HUNTINGTON, Ellsworth, *Civilization and Climate*. New Haven: Yale University Press, 1915.

Hurricane. Washington, D.C.: U.S. Department of Commerce, 1971.

HUXLEY, Aldous, *On Art and Artists*, Morris Philipson, ed. Cleveland: Meridian Books, 1960.

————, *Collected Essays*. New York: Harper & Bros., 1959; Bantam, 1960.

————, *Tomorrow and Tomorrow and Tomorrow*. New York: Harper & Row, 1956; Signet, 1964.

"Ice Ages and the Galaxy's Spiral Arms," *Science News*, July 12, 1975; errors noted in letters, *Science News*, August 16, 1975.

"An Icy Alaska Delay," *Time*, October 6, 1975.

IDSO, Sherwood B., "Thermal Blanketing: A Case for Aerosol-Induced Climatic Alteration," *Science*, October 4, 1974.

"Israel Possesses . . . Up to 60,000 Tons of Uranium . . . ," *Los Angeles Times*, October 9, 1975.

IDYLL, C. P., "The Anchovy Crisis," *Scientific American*, June 1973.

————, *The Sea Against Hunger*. New York: Crowell, 1970.

INGHAM, M. F., "The Spectrum of the Airglow," *Scientific American*, January 1972.

INGSTAD, Helge, *Westward to Vinland*. New York: St. Martin's Press, 1969.

"Interplanetary and Terrestrial Magnetism," *Science News*, October 16, 1971.

IRVING, Laurence, "Adaptations to Cold," *Scientific American*, January 1966.

"Is Climate Influenced by Earth's Magnetism?" *Science News*, February 2, 1974.

"Is Gravity Weakening?" *Science News*, August 24 & 31, 1974.

JAYAWEERA, K. O. L. F., and Takeshi OHTAKE, "Artificial Cloud Formation in the Atmosphere," *Science*, November 3, 1972.

JENNINGS, Peter R., "Rice Breeding and World Food Production," *Science*, December 20, 1974.

JOHNSON, Stanley, *The Green Revolution*. New York: Harper & Row, 1972.

JONES, Robert A., "Genetic Erosion of Plants Poses Threat to Crops," *Los Angeles Times*, June 4, 1974.

JOSEPH, J. H., et al., "Desert Aerosols Transported by Khamsinic Depressions and Their Climatic Effects," *Journal of Applied Meteorology*, August 1973.

"The Jupiter-Effect Effect," *Science News*, December 13, 1975.

KAHN, Rasheeduddin, "Is India Governable?" *Seminar* (India), January 1975; *Atlas World Press Review*, September 1975.

KAPLAN, Lewis D., "Climate Control—Dream or Nightmare?" *New Scientist*, March 9, 1961; *Science Digest*, July 1961.

KAULA, William M., "Earth, Gravitational Field of," *Encyclopedia Britannica*, Macropedia, Vol. 6, 1974.

KAVALER, Lucy, *Freezing Point*. New York: John Day, 1970.

KELLOGG, W. W., and S. H. SCHNEIDER, "Climate Stabilization: For Better or for Worse?" *Science*, Vol. 186, pp. 1163-72, Fig. 4, December 27, 1974.

KENDREW, W. G., *The Climates of the Continents*, 4th ed. London: Oxford University Press, 1953.

KENNAN, George F., "Reappraising Our Vital Interests," *Foreign Affairs*, October 1972.

"Kennedy Hopeful on Russian Amity," *New York Times*, November 4, 1960.

KENNETT, James P., and Robert C. THUNELL, "Global Increase In Quaternary Explosive Volcanism," *Science*, February 14, 1975.

KHARKOVSKY, Alexander, "Neuston, A Living Continent," *Soviet Life*, July 1974.

KIMBLE, George H. T., "The Changing Climate," *Scientific American*, April 1950.

KING, J. W., "Sun-Weather Relationships," *Astronautics & Aeronautics*, April 1975.

KLASS, Philip J., "Laser Weapons—#3—Current Systems Still More Cost-Effective," *Aviation Week & Space Technology*, September 8, 1975.

KRAUSKOPF, Konrad B., and Arthur BEISER, *The Physical Universe*, 3rd ed. New York: McGraw-Hill, 1973.

KRENKEL, Peter A., and Frank L. PARKER, eds., *Biological Aspects of Thermal Pollution*. Nashville, Tenn.: Vanderbilt University Press, 1969.

KRICK, Irving P., and Roscoe FLEMING, *Sun, Sea, and Sky*. Philadelphia: Lippincott, 1954.

KROOK, Max, "Interstellar Matter and the Solar Constant," in Harlow SHAPLEY, ed., *Climatic Change*. Cambridge: Harvard University Press, 1953.

KUKLA, George J., and Helena J. KUKLA, "Increased Surface Albedo in the Northern Hemisphere," *Science*, February 22, 1974.

KUKLA, G. J., and R. K. MATTHEWS, "When Will the Present Interglacial End?" *Science*, October 13, 1972.

KURTÉN, Björn, "Continental Drift and Evolution," *Scientific American*, March 1969.

——, *The Ice Age*. New York: Putnam, 1972.

LAFFIN, John, *The Hunger to Come*, rev. ed. London: Abelard-Schuman, 1971.

LAMB, Hubert H., *The Changing Climate*. London: Methuen, 1966.

——, "Climate," in *Encyclopedia Britannica*, Macropedia, Vol. 4, 1974.

——, *Climate: Present, Past and Future*, Vol. I. London: Methuen, 1972.

——, "Is The Earth's Climate Changing?" *UNESCO Courier*, August–September 1973.

LANDSBERG, Helmut E., "Man-Made Climatic Changes" *Science*, December 18, 1970.

LANGER, William L., "Checks on Population Growth: 1750–1850," *Scientific American*, February 1972.

LANGWAY, Lynn, and Dewey GRAM, "Grain: The Big Five," *Newsweek*, August 4, 1975.

LANSFORD, Henry, "Climate Outlook: Variable and Possibly Cooler," *Smithsonian*, November 1975.

LAPORTE, Léo F., *Ancient Environments*. Englewood Cliffs, N.J.: Prentice-Hall, 1968.

LAPP, Ralph E., "We May Find Ourselves Short of Uranium, Too," *Fortune*, October 1975.

LAPPÉ, Frances Moore, "Fantasies of Famine," *Harper's Magazine*, February 1975.

LASAGA, Antonio C., et al. "Primordial Oil Slick," *Science*, October 1, 1971.

"Latitude Gradient in the Solar Wind," *Science News*, April 12, 1975.

LE ROY LADURIE, Emmanuel, *Times of Feast, Times of Famine*. New York: Doubleday, 1971.

"Less Sunshine In the United States," *Science News*, April 12, 1975; similar story, *New York Times*, March 20, 1975.

LEVITSKY, Serge L., "Russia's Fantastic Weather Plan," *Coronet*, June 1959.

LEWALLEN, John, *Ecology of Devastation: Indochina*. Baltimore: Penguin, 1971.

LEWIN, Roger, "Starved Brains," *Psychology Today*, September 1975.

LEWIS, John P., *Quiet Crisis in India*. Washington, D.C.: Brookings Institution, 1962; Anchor Books, 1964.

LIKENS, Gene E., and F. Herbert BORMANN, "Acid Rain . . . ," *Science*, June 14, 1974.

LINEBERRY, William P., ed., *Priorities for Survival*. New York: H. W. Wilson, 1973.

LIVINGSTON, William C., "Magnetic Fields on the Quiet Sun," *Scientific American*, November 1966.

LIVINGSTONE, Daniel A., "Speculations on the Climatic History of Mankind," *American Scientist*, May 1971.

LOGSDON, Gene, "Grow Your Own Nitrogen Factory," *Organic Gardening and Farming*, March 1975.

LONG, Capt. E. John, *Ocean Sciences*. Annapolis: United States Naval Institute, 1964.

LORENZ, Edward N., "The Circulation of the Atmosphere," *Scientific American*, December 1966.

——, "Climatic Change as a Mathematical Problem," in William H. MATTHEWS, et al., eds., *Man's Impact on Climate*. Cambridge: MIT Press, 1971.

LOVINS, Amory Block, "The Energy Problems in our Future," *Ceres*, November–December 1974.

LOWDERMILK, W. C., "The Reclamation of a Man-Made Desert," *Scientific American*, March 1960.

LOWRY, William P., "The Climate of Cities," *Scientific American*, August 1967.

——, *Weather and Life*. New York: Academic Press, 1969.

"Lunar Dust, Earthly Ice and the Galaxy," *Science News*, November 15, 1975.

MacDONALD, Gordon J. F., "How to Wreck the Environment," in Nigel CALDER, ed., *Unless Peace Comes*. New York: Viking, 1968.

MacINTYRE, Ferren, "The Top Millimeter of the Ocean," *Scientific American*, May 1974.

——, "Why The Sea Is Salt," *Scientific American*, November 1970.

McCARTHY, Raymond L., "Why Industry Urges Delay On Spray Ban," *Los Angeles Times*, April 27, 1975.

McCARTHY, Richard D., *The Ultimate Folly*. New York: Knopf, 1969.

McCARTHY, Terence, "Feast or Famine: The Choices for Mankind," *Ramparts*, September 1974.

McDONALD, James E., "The Coriolis Effect," *Scientific American*, May 1952.

McKEE, Alexander, *Farming the Sea*. New York: Crowell, 1969.

McLARNEY, William, "Why Not Carp?" *Organic Gardening*, February 1972.

McLAUGHLIN, Robert H., "They Stole Our Rain," *Vista*, August 1973.
McNEIL, Mary, "Lateritic Soils," *Scientific American*, November 1964.
MADDOX, John, *Beyond the Energy Crisis*. New York: McGraw-Hill, 1975.
"Magnetic Reversals: Effects on Life . . . ," *Science News*, April 6, 1974.
"Mainland U.S. Volcano Could Erupt," *Science News*, March 8, 1975.
MALLOY, Michael T., "If the 'New Order' Wins . . . We'll Pay the Bill," *National Observer*, August 30, 1975.
_____, "The Next Crisis: Universal Famine," *National Observer*, March 30, 1974.
_____, "Soviet 'Ocean of Grain' Isn't Enough," *National Observer*, August 23, 1975.
"Man Heats His Planet," *Science News*, July 7, 1973.
"Man's Impact On Climate: What Is Ahead?" *Science News*, July 31, 1971.
MANCKE, Richard B., *The Failure of U.S. Energy Policy*. New York: Columbia University Press, 1974.
MARKSON, Ralph, "Atmospheric Electrical Detection of Organized Convection," *Science*, June 20, 1975.
MARSHALL, Burke (panel chairman), et al., *Safeguarding the Atom*. New York: United Nations Association, 1972.
MARX, Wesley, *Man and His Environment: Waste*. New York: Harper & Row, 1971.
MATTHEW, William D., *Climate and Evolution*. New York: Arno Press, 1974.
MATTHEWS, Samuel W., "This Changing Earth," *National Geographic*, January 1973.
MATTHEWS, William H., et al., eds., *Man's Impact on Climate*. Cambridge: MIT Press, 1971.
"Maybe the Atmosphere Hasn't Changed," *Science News*, September 8, 1973.
MAYER, Jean, "What We Must Do to Feed the World," *Reader's Digest*, September 1975.
_____, "*Why* the Food Crisis?" *Reader's Digest*, May 1975.
MEADOWS, D. L., and D. H. MEADOWS, *Toward a Global Equilibrium*. Cambridge, Mass.: Wright-Allen, 1974.
MEADOWS, D. L., et al., *Dynamics of Growth in a Finite World*. Cambridge, Mass.: Wright-Allen, 1974.
MEADOWS, Donella H., et al., *The Limits to Growth*. New York: Universe Books, 1972.
MEDVIN, Norman, *The Energy Cartel*. New York: Vintage, 1974.
MEINEL, Aden and Marjorie, "Stratospheric Dust-Aerosol Event of November 1974," *Science*, May 2, 1975.
MEISLER, Stanley, "On Dividing the Seas: History Made Waves," *Los Angeles Times*, June 20, 1974.
MENDES, Lucas, "The Mysterious Deaths of French Nuclear Scientists" (interview with Lowell Ponte), *Manchete* (Brazil), June 2, 1973.
MENZEL, Donald H., "On the Causes of the Ice Ages," in Harlow SHAPLEY, ed., *Climatic Change*. Cambridge: Harvard University Press, 1953.
MESAROVIC, Mihajlo, Eduard PESTEL, and Maurice GUERNIER, "A Computer Warning of Hunger Tomorrow," *UNESCO Courier*, July–August 1974.
MILES, Marvin, "Scientists Test Plan for Orbital Power Beams," *Los Angeles Times*, October 6, 1975.
"Military Rainmaking Confirmed by U.S.," *Science News*, May 25, 1974.
MILLIKAN, N. F., and David HAPGOOD, *No Easy Harvest: The Dilemma of Agriculture in Underdeveloped Countries*. Boston: Little, Brown, 1967.
"Missing Neutrinos: A Competing Force?" *Science News*, July 12, 1975.
MITCHELL, J. Murray, Jr., "The Problem of Climatic Change and Its Causes," in

William H. MATTHEWS, et al., eds., *Man's Impact on Climate.* Cambridge: MIT Press, 1971.

MIYAKODA, Kikuro, "Numerical Weather Prediction," *American Scientist,* September–October 1974.

MONIN, Andrei, "Joint Exploration of the Ocean/Cooperation on the 180th Meridian," *Soviet Life,* July 1974.

MORDY, Wendell, "Toward a World Atmospheric Community," *Center Report,* February 1974.

"More to This than Meets the Ozone," *Science News,* January 18, 1975.

MOSLEY, Leonard, *Power Play: Oil in the Middle East.* New York: Random House, 1973.

MOYNIHAN, Daniel P., "The United States and the New World Society," *Commentary,* March 1975; *Reader's Digest,* June 1975.

MURPHY, Earl Finbar, *Governing Nature.* Chicago: Quadrangle, 1968.

———, *Man and his Environment: Law.* New York: Harper & Row, 1971.

MURRAY, Bruce C., *Navigating the Future.* New York: Harper & Row, 1975.

MUSTEL, E. R., *On the Reality of the Influence of Solar Corpuscular Streams upon the Lower Layer of the Earth's Atmosphere.* Publication No. 24, Astronomical Council, U.S.S.R. Academy of Sciences, Moscow, 1972.

MYRDAL, Gunnar, *The Challenge of World Poverty.* New York: Pantheon, 1970.

———, "Political Factors in Economic Assistance," *Scientific American,* April 1972.

———, "The Transfer of Technology to Underdeveloped Countries," *Scientific American,* September 1974.

MYRDAL, Jan, and Gun KESSLE, *Angkor.* New York: Random House, 1970; Vintage, 1971.

NAMIAS, Jerome, "Climatic Anomaly Over the United States During the 1960's," *Science,* November 13, 1970.

NAMIAS, Jerome, and Joseph CHI KAN HUANG, "Sea Level at Southern California: A Decadal Fluctuation," *Science,* July 28, 1972.

"NAS Warning on Climate Changes," *Science News,* January 25, 1975.

National Academy of Sciences/NRC, *The Earth and Human Affairs.* San Francisco: Canfield Press, 1972.

———, *Plan for U.S. Participation in the Global Atmospheric Research Program.* Washington, D.C.: 1975.

———, *Understanding Climatic Change: A Program for Action.* Washington, D.C.: 1975.

———, *Weather and Climate Modification: Problems and Progress.* Washington, D.C.: 1973.

———, *Weather and Climate Modification: Problems and Prospects.* Washington, D.C.: 1966.

NAWROCKI, Paul J., and Robert PAPA, *Atmospheric Processes.* Englewood Cliffs, N.J.: Prentice-Hall, 1963.

NEIBURGER, Morris, et al., *Understanding Our Atmospheric Environment.* San Francisco: Freeman, 1973.

NEILANDS, J. B., et al., *Harvest of Death: Chemical Warfare in Vietnam and Cambodia.* New York: Free Press, 1972.

NELSON, Bryce, "Drive Launched to Save Great Plains Windbreak Trees," *Los Angeles Times,* August 28, 1975.

———, "Physicist Says Bering Strait Dam Would Bless Southern California," *Los Angeles Times,* January 27, 1973.

"New 'Anomalies' in Climates Seen," *New York Times,* February 16, 1975.

NEWELL, Reginald E., "The Amazon Forest and Atmospheric General Circulation," in William H. MATTHEWS, et al., eds., *Man's Impact on Climate*. Cambridge: MIT Press, 1971.
————, "The Circulation of the Upper Atmosphere," *Scientific American*, March 1964.
————, "The Global Circulation of Atmospheric Pollutants," *Scientific American*, January 1971.
NEWMAN, J. E., and R. C. PICKETT, "World Climates and Food Supply Variations," *Science*, December 6, 1974.
NISHIOKA, Hideo, *Long Warm Weather Cycle Ahead*. New York: International Economic Research Bureau, 1957.
"Now, the Bee Shortage," *Newsweek*, November 11, 1974.
"Ocean-floor Record Links Dinosaurs, Plankton, Climate," *Science News*, October 23, 1971.
"Oceans Approach Man as CO Producer," *Science News*, September 8, 1973.
ODELL, Peter, *Oil and World Power*. Baltimore: Penguin, 1972.
ODUM, Howard T., *Environment, Power and Society*. New York: Wiley, 1971.
"Official Warns of Terrorist A-Arms," *Los Angeles Times*, September 22, 1975.
"Of Man and Milk," *Time*, July 13, 1970.
OGNIBENE, Peter J., "Making War With the Weather," *New Republic*, September 30, 1972.
"Oil-Eating Bug," *Time*, September 22, 1975.
OLIVER, John E., *Climate and Man's Environment*. New York: Wiley, 1973.
O'NEILL, Gerard K., "Space Colonies and Energy Supply to Earth," *Science*, December 5, 1975.
————, "The Colonization of Space," *Physics Today*, September 1974.
OPHULS, William, "The Scarcity Society," *Harper's Magazine*, April 1974.
ÖPIK, Ernest J., "Climate and the Changing Sun," *Scientific American*, June 1958.
ORDWAY, Richard J., *Earth Sciences*, 2nd ed. New York: Van Nostrand Reinhold, 1972.
ORGANSKI, A. F. K., and Katherine ORGANSKI, *Population and World Power*. New York: Knopf, 1961.
ORVILLE, Howard T., "Weather as a Weapon," *Popular Science*, June 1958.
ORVILLE, Howard T., et al., *Final Report of the Advisory Committee on Weather Control*, Vols. I & II. Washington, D.C.: U.S. Government Printing Office, 1957.
OSTRANDER, Sheila, and Lynn SCHROEDER, *Psychic Discoveries Behind the Iron Curtain*. Englewood Cliffs, N.J.: Prentice-Hall, 1970.
O'TOOLE, Thomas, "Study of Climate Urged as Nation's Main Space Goal." *Los Angeles Times*, February 22, 1976.
OTTERMAN, J., "Baring High-Albedo Soils by Overgrazing: A Hypothesized Desertification Mechanism," *Science*, November 8, 1974.
"Overgrazing May Cause Droughts," *Los Angeles Times*, December 15, 1975.
OXENSTIERNA, Eric, "The Vikings," *Scientific American*, May 1967.
PADDOCK, William, and Paul PADDOCK, *Famine 1975!* Boston: Little, Brown, 1967.
PALLOTTELLI, Duilio, "U.S. Laser Weapons Development" (interview with Lowell Ponte), *L'Europeo* (Italy), January 18, 1973.
PARKER, E. N., "The Sun," *Scientific American*, September 1975.
PAULY, K. A., "Cause of the Great Ice Ages," *Scientific Monthly*, August 1952.
PEIXOTO, José P., and M. Ali KETTANI, "The Control of the Water Cycle," *Scientific American*, April 1973.
PELL, Claiborne, "End Weather Warfare Before It's Too Late," *Los Angeles Times*, October 8, 1973.

_____, "Prohibition Urged on Use of Environmental Modification for Military Purposes," *Congressional Record* (Senate), August 22, 1974.

_____, "U.S. Abstention at the U.S. General Assembly," *Congressional Record* (Senate), December 19, 1974.

"Pentagon: Weather as a Weapon of War," *New York Times*, July 9, 1972.

PERRY, J. P., "The Special Case of Chemical and Biological Weapons," *Bulletin of Atomic Scientists*," May 1975.

PETERSEN, W., *The Politics of Population*. New York: Doubleday, 1964.

PETERSON, James T., "Energy and the Weather," *Environment*, October 1973.

PETERSON, James T., and Christian E. JUNGE, "Sources of Particulate Matter in the Atmosphere," in William H. MATTHEWS, et al., eds., *Man's Impact on Climate*. Cambridge: MIT Press, 1971.

PIERCE, A. Keith, and Edith A. MULLER, "Sun," *Encyclopedia Britannica*, Macropedia, Vol. 17, 1974.

PIMENTEL, David, et al., "Food Production and the Energy Crisis," *Science*, November 2, 1973.

PIRAGES, Dennis C., and Paul R. EHRLICH, *ARK II*. New York: Viking Press, 1974.

PIRIE, N. W., *Food Resources Conventional and Novel*. Baltimore: Penguin Books, 1969.

PLASS, Gilbert N., "Carbon Dioxide and Climate," *Scientific American*, July 1959.

POLEMAN, Thomas T., "World Food: A Perspective," *Science*, May 9, 1975.

POLLACK, James B., "Mars," *Scientific American*, September 1975.

PONTE, Lowell, "Better Do As We Say; This Is An Atom Bomb . . . ," *Penthouse*, February 1972.

_____, "The Body Electric," *Penthouse*, February 1974.

_____, "The Danger of Terrorists Getting Illicit A-bombs," *Los Angeles Times*, January 15, 1973.

_____, "Dangers of a National Power Breakdown," *Los Angeles Times*, November 9, 1973.

_____, "The Dangers of Technology Addiction," *Skeptic*, Special Issue #6—*America's Survival*, March–April 1975.

_____, "Deadly U.S. Arsenal in a Real War of Nerves," *Los Angeles Times*, August 10, 1972.

_____, "An Ice Age Cometh?" *East-West Journal*, January 1975.

_____, "Manipulation, Morality and Weather Mod," *Skeptic*, Special Issue #2—*Scarcity*, July–August 1974.

_____, *The New Cold War: American and Soviet Ideas for Dealing With the World's Cooling Climate* (monograph). Los Angeles: 1974.

_____, "Nippon Goes Nuclear," *The Progressive*, September 1973; excerpts in *Friday Review of Defense Literature*, U.S. Department of Defense, September 7, 1973.

_____, "Pearl Harbor: Japan's Reaction to an Oil Embargo," *Los Angeles Times*, December 7, 1973.

_____, "The Proliferation Non-Treaty," *National Review*, December 15, 1970.

_____, "SABMIS" (discussion comments), *U.S. Naval Institute Proceedings*, November 1969.

_____, "Terrorists Might Even Go Nuclear," *Los Angeles Times*, June 9, 1974.

_____, "Tomorrow The Terrorists May Hijack New York City," *Gallery*, February 1973.

_____, "U.S. Studies Ozone Layer Weapons" (syndicated column), *Anaheim Bulletin*, June 21, 1972.

_____, "Weather Warfare Forecast: Partly Cloudy," *Los Angeles Times*, January 29, 1976.

POTTER, Neal, *Natural Resource Potentials of the Antarctic.* New York: American Geographical Society, 1969.
"Prehistoric Flood from Ice Surge," *Science News,* October 4, 1975.
PROBSTEIN, Ronald F., "Desalination," *American Scientist,* May 1973.
"Protein Research: Helping the Hungry," *Science News,* September 13, 1975.
PURRETT, Louise A., "Analyzing the Atmosphere," *Science News,* July 22, 1972.
———, "The Shifting World of Arcti Sea Ice," *Science News,* July 31, 1971.
———, "The Tropics Weren't So 'Stable' After All," *Science News,* September 11, 1971.
———, "Weather Modification as a Future Weapon," *Science News,* April 15, 1972.
"Pushing the Arab Cause in America," *Time,* June 23, 1975.
"Question for the (Ice) Ages," *Science News,* July 17, 1971.
"The Race With Hunger," *Newsweek,* November 11, 1974.
RADFORD, Jeff, "Air Pollution Linked to Floods, Drought," *Christian Science Monitor,* August 27, 1974.
The Radio Amateur's Handbook, 52nd ed. Newington, Conn.: American Radio Relay League, 1975.
RAIKES, Robert, *Water, Weather, and Prehistory.* London: John Baker, 1967.
RAMANATHAN, Veerabhadran, "Greenhouse Effect Due to Chlorofluorocarbons: Climatic Implications," *Science,* October 3, 1975.
RAND, Christopher T., *Keeping Democracy Safe for Oil.* Boston: Little, Brown, 1975.
RAND Weather Modification Research Project, *Weather Modification Progress and the Need for Interactive Research.* Santa Monica, Cal.: RAND Corporation, 1968.
RANK, D. M., et al., "Interstellar Molecules and Dense Clouds," *Science,* December 10, 1971.
RASOOL, S. I., ed., *Chemistry of the Lower Atmosphere.* New York: Plenum, 1973.
RASOOL, S. I., and S. H. SCHNEIDER, "Atmospheric Carbon Dioxide and Aerosols: Effects of Large Increases on Global Climate," *Science,* July 9, 1971; letters in response, September 10, 1971.
RATCLIFFE, J. A., *Sun, Earth and Radio.* New York: McGraw-Hill, 1970.
REICHLEY, A. James, "The Case for Interdependence," *Fortune,* April 1975.
REITER, Elmar R., *Jet Streams.* New York: Doubleday, 1967.
REVELLE, Roger, "Can The Earth Feed the Growing Multitudes?" *UNESCO Courier,* July–August 1974.
———, "Food and Population," *Scientific American,* September 1974.
———, "The Role of the Oceans," *Saturday Review,* May 7, 1966.
RIDGEWAY, James, *The Last Play.* New York: Dutton, 1973.
———, *New Energy.* Boston: Beacon, 1975.
RIENOW, Robert, and Leona Train RIENOW, *Moment in the Sun.* New York: Dial, 1967.
RINEHART, John S., "18.6-Year Earth Tide Regulates Geyser Activity," *Science,* July 28, 1972.
RISEBROUGH, R. W., et al., "Pesticides: Transatlantic Movements in the Northeast Trades," in Arthur S. BOUGHEY, ed., *Readings in Man, the Environment, and Human Ecology.* New York: Macmillan, 1973.
RITCHIE, J. D., "Use Legumes to Convert to Natural Nitrogen," *Organic Gardening and Farming,* March 1975.
ROBERTS, Walter Orr, "Climate Control," *Physics Today,* August 1967.
———, "Man on a Changing Earth," *American Scientist,* January 1971.
ROCKS, Lawrence, and Richard P. RUNYON, *The Energy Crisis.* New York: Crown, 1972.
RODALE, Robert, "Famine in America is Possible," *Organic Gardening and Farming,*

September 1974.

———, "Human Wastes Can Be Recycled," *Organic Gardening and Farming*, February 1972.

———, "Japan's New Approaches to Organic Methods," *Organic Gardening and Farming*, March 1973.

———, "Learning from India," *Organic Gardening and Farming*, June 1975.

———, "Living on the Land in China Today," *Organic Gardening and Farming*, May 1973.

———, *Sane Living in a Mad World*. Emmaus, Pa: Rodale Press, 1972; New York: Signet, 1973.

———, "Walking Through China's Garden," *Organic Gardening and Farming*, June 1973.

ROSENBERG, Norman J., *Microclimate*. New York: Wiley, 1974.

ROSENFELD, Albert, *The Second Genesis*. Englewood Cliffs, N.J.: Prentice-Hall, 1969.

ROSENZWEIG, Michael L., *And Replenish the Earth*. New York: Harper & Row, 1974.

ROWLAND, F. Sherwood, "Danger to Ozone: Evidence Supports Immediate Action," *Los Angeles Times*, April 27, 1975.

RUSSELL, Cristine, "Weather Control: Tool or Weapon?" *Santa Monica Evening Outlook*, August 30, 1975.

"Russia-Alaska Tunnel Visioned by Red Scientists," *Los Angeles Times*, February 2, 1946.

RUTTAN, Vernon W., et al., eds., *Agricultural Policy in an Affluent Society*. New York: Norton, 1969.

RYDER, Norman B., "The Family in Developed Countries," *Scientific American*, September 1974.

SAFRANY, David R., "Nitrogen Fixation," *Scientific American*, October 1974.

SAGAN, Carl, *The Cosmic Connection*. New York: Doubleday, 1973; Dell, 1975.

SAGAN, Carl, Jonathan Norton LEONARD, et al., *Planets*. New York: Time-Life Books, 1966.

"Sahara's Far-flung Dust Dims Atlantic Sunlight," *Science News*, January 19, 1974.

SALINGER, M. J., "Climate Change in New Zealand" (letter), *Science News*, September 27, 1975.

SALISBURY, David F., "Cities in Space . . . ," *Christian Science Monitor*, August 22, 1975.

———, "Major Crop Failures Foreseen," *Christian Science Monitor*, August 27, 1974; *Current*, October 1974.

SALKELD, Robert, "Space Colonization Now?" *Astronautics & Aeronautics*, September 1975.

SANCHEZ, P. A. and S. W. BUOL, "Soils of the Tropics and the World Food Crisis," *Science*, May 9, 1975.

SANDERSON, Fred H., "The Great Food Fumble," *Science*, May 9, 1975.

"Sardines & Hurricanes," *Time*, February 7, 1955.

"Saving the Caspian," *Newsweek*, March 17, 1975.

SCHAEFER, Vincent J., "The Inadvertent Modification of the Atmosphere by Air Pollution," in Arthur S. BOUGHEY, ed., *Readings in Man, the Environment, and Human Ecology*. New York: Macmillan, 1973.

SCHINDLER, D. W., et al., "Atmospheric Carbon Dioxide: Its Role in Maintaining Phytoplankton Standing Crops," *Science*, September 29, 1972.

SCHNEIDER, Barry, "Big Bangs from Little Bombs," *Bulletin of Atomic Scientists*, May 1975.

SCHNEIDER, Stephen H., *The Genesis Strategy*. New York: Plenum, 1976.

———, "On the Carbon Dioxide-Climate Confusion," *Journal of the Atmospheric Sciences*, November 1975.

SCHNEIDER, Stephen H., and Roger D. DENNETT, "Climatic Barriers to Long-Term Energy Growth," *Ambio*, 1975. Vol. 4, No. 2.

SCHNEIDER, Stephen H., and Clifford MASS, "Volcanic Dust, Sunspots, and Temperature Trends," *Science*, November 21, 1975.

SCHULTZ, Gwen, *Glaciers and the Ice Age*. New York: Holt, Rinehart and Winston, 1963.

———, *Ice Age Lost*. New York: Doubleday, 1974.

SCHUMACHER, E. F., *Small Is Beautiful*. New York: Harper & Row, 1973.

SCHWARTZ, Eugene S., *Overskill*. Chicago: Quadrangle, 1971.

SCOTT, William D., and Zev LEVIN, "Open Channels in Sea Ice (Leads) As Ion Sources," *Science*, August 4, 1972.

"Sea-Air Explanation," *Science News*, November 28, 1970.

SEABORG, Glenn T., and Justin L. BLOOM, "Fast Breeder Reactors," *Scientific American*, November 1970.

SEARS, Paul B., "Climate and Civilization," in Harlow SHAPLEY, ed., *Climatic Change*. Cambridge: Harvard University Press, 1953.

———, "Climate and Culture—New Evidence," *Science*, July 13, 1951.

———, *Deserts on the March*, 3rd rev. ed. Norman, Okla.: University of Oklahoma Press, 1967.

———, *Where There Is Life*, rev. ed. New York: Dell, 1970.

SELLERS, W. D., *Physical Climatology*. Chicago: University of Chicago Press, 1965.

———, "Reassessment of the Effect of CO Variations on a Simple Global Climatic Model," *Journal of Applied Meteorology*, October 1974.

SEWELL, W. R. Derrick, ed., *Human Dimensions of Weather Modification*. Chicago: University of Chicago Department of Geography Research Series No. 100, 1965; University of Chicago Press, 1966.

———, "Weather Modification: When Should We Do It and How Far Should We Go?" in Robert G. FLEAGLE, ed., *Weather Modification*. Seattle: University of Washington Press, 1969.

SHAPLEY, Deborah, "Plutonium: Reactor Proliferation Threatens a Nuclear Black Market," *Science*, April 9, 1971.

———, "Rainmaking: Rumored Use Over Laos Alarms Arms Experts, Scientists," *Science*, June 16, 1972.

———, "Rainmaking: Stockholm Stand Watered Down for Military, *Science*, June 30, 1972.

———, "Technology in Vietnam: Fire Storm Project Fizzled Out," *Science*, July 21, 1972.

———, "Weather Warfare: Pentagon Concedes 7-Year Vietnam Effort," *Science*, June 7, 1974.

SHAPLEY, Harlow, ed., *Climatic Change*. Cambridge: Harvard University Press, 1953.

SHAW, David, "Sunspots and Temperature," *Journal of Geophysical Research*, October 15, 1965.

SIEVER, Raymond, "The Earth," *Scientific American*, September 1975.

SILVERBERG, Robert, *The Challenge of Climate*. New York: Meredith Press, 1969.

———, *Lost Cities and Vanished Civilizations*. Radnor, Pa.: Chilton, 1962; New York: Bantam, 1974.

SIMON, Paul and Arthur, *The Politics of World Hunger*. New York: Harper's Magazine Press, 1973.

SINGER, S. F., ed., *Global Effects of Environmental Pollution*. New York: Springer-Verlag, 1970.

SKINNER, Brian J., and Karl K. TUREKIAN, *Man and the Ocean*. Englewood Cliffs, N.J.: Prentice-Hall, 1973.

SKOLNIKOFF, Eugene B., *Science, Technology, and American Foreign Policy*. Cambridge: MIT Press, 1967.

"Skylab Rocket Reportedly Tore Hole in Ionosphere," *Los Angeles Times*, May 12, 1975.

SMAGORINSKY, Joseph, "Large Scale Atmospheric Circulation," in William H. MATTHEWS, et al., eds., *Man's Impact on Climate*. Cambridge: MIT Press, 1971.

SMITH, Audrey U., "Life at Low Temperatures," *Nature*, October 4, 1958.

SMITH, F. G. Walton, *The Seas in Motion*. New York: Crowell, 1973.

SMITH, L. P., "Life in a Changing Climate," in Nigel CALDER, ed., *Nature In The Round*. New York: Viking, 1973.

"Solar Neutrinos: Change the Theory," *Science News*, April 27, 1974.

SOLHEIM, Wilhelm G. II, "An Earlier Agricultural Revolution," *Scientific American*, April 1972.

"Something Is Wrong With the Weather," (interview with Dr. J. Murray MITCHELL, Jr.), *U.S. News & World Report*, July 10, 1967.

"Something New Under The Sun: A Flicker," *Science News*, July 19, 1975.

"Soviet Scientist Offers Perpetual Summer Idea," *New York Times*, March 13, 1960.

"Soviets in U.N. Decry Weather Warfare," *Science News*, November 2, 1974.

"Space Colonies: Home, Home on Lagrange," *Science News*, September 6, 1975.

SPOFFORD, Walter O., Jr., "Decision Making Under Uncertainty: The Case of Carbon Dioxide Buildup In The Atmosphere," in William H. MATTHEWS, et al., eds., *Man's Impact on Climate*. Cambridge: MIT Press, 1971.

SPRAGUE, G. F., "Agriculture in China," *Science*, May 9, 1975.

SRIDHARAN, Sumi, "The Green Revolution Runs Down," *Hindustan Times*, March 20, 1975; *Atlas World Press Review*, May 1975.

STARR, Thomas B., *A Survey of the State of the Art in Meteorological Long-Range Statistical Forecasting*. Madison, Wis.: Institute for Environmental Studies, 1974.

STARR, Victor P., and Peter A. GILMAN, "The Circulation of the Sun's Atmosphere," *Scientific American*, January 1968.

STEFANSSON, Vihjalmur, *The Northward Course of Empire*. New York: Macmillan, 1924.

STEVENS, Leonard A., "The County That Reclaims Its Sewage," *Reader's Digest*, July 1975.

STEWART, J. Clayton, "Cold Can Do You In," *Sports Illustrated*, March 10, 1975; *Reader's Digest*, October 1975.

STEWART, R. W., "The Atmosphere and the Ocean," *Scientific American*, September 1969.

STILL, Henry, *The Dirty Animal*. New York: Hawthorn, 1967.

STINE, G. Harry, *The Third Industrial Revolution*. New York: Putnam, 1975.

STOMMEL, Henry, "The Circulation of the Abyss," *Scientific American*, July 1958.

———, *The Gulf Stream*, 2nd ed. Berkeley, Cal.: University of California Press, 1966.

STONIER, Tom, "Ecological Upsets: Climate and Erosion," in Cecil E. JOHNSON, ed., *Eco-Crisis*. New York: Wiley, 1970.

STRINGER, E. T., *Foundations of Climatology*. San Francisco: Freeman, 1972.

———, *Techniques of Climatology. San Francisco;* Freeman, 1972.

STRONG, Maurice, ed., *Who Speaks for Earth?* New York: Norton, 1973.

SULAKVELIDZE, G. K., et al., "Progress of Hail Suppression Work in the U.S.S.R.," in W. N. HESS, ed., *Weather and Climate Modification*. New York: Wiley, 1974.

SULLIVAN, Walter, "Bering Dam Held Disastrous Plan," *New York Times*, November 15, 1960.

——, "Cosmic Ray Cycle Poses Mystery," *New York Times*, March 13, 1960.

——, "Plan for [Bering] Dam is Old," *New York Times*, October 25, 1959.

——, "Scientists Ask Why World Climate Is Changing; Major Cooling May Be Ahead," *New York Times*, May 21. 1975.

——, "Two Climate Experts Decry Predictions of Disasters," *The New York Times*, February 22, 1976.

——, "Warming Trend Seen in Climate," *New York Times*, August 14, 1975.

"Sun, Caves and Ice Ages," *Science News*, May 25, 1974.

"Sun Controls Weather," *Science News Letter*, October 15, 1955.

"Sunlight and the SST," *Saturday Review*, April 1973.

"Sun's Magnetism and Earth's Winds," *Science News*, June 1, 1974.

SUOMI, Verner E., "GARP: Dreams into Realities," *Astronautics & Aeronautics*, October 1975.

SUTCLIFFE, R. C., *Weather and Climate*. New York: Norton, 1966.

SUTTON, O. G., *The Challenge of the Atmosphere*. New York; Harper & Bros., 1961.

SVALGAARD, Leif, and John M. WILCOX, "The Spiral Interplanetary Magnetic Field: A Polarity and Sunspot Cycle Variation," *Science*, October 4, 1974.

TAKEUCHI, H., et al., *Debate About The Earth*, rev. ed. San Francisco: Freeman, Cooper, 1970.

TAMLIN, Arthur R., and John W. GOFMAN, *"Population Control" Through Nuclear Pollution*. Chicago: Nelson-Hall, 1970.

TANZER, Michael, *The Political Economy of International Oil and the Underdeveloped Countries*. Boston: Beacon, 1969.

TAUBENFELD, Howard J., *Controlling the Weather*. Port Washington, N.Y.: Dunellen, 1970.

——, *Weather Modification Law, Controls, Operations*. Washington, D.C.: National Science Foundation, 1966.

——, ed., *Weather Modification and the Law*. Dobbs Ferry, N.Y.: Oceana Publications, 1968.

TAYLOR, George C., "Water, History, and the Indus Plain," *Natural History*, May 1965.

TAYLOR, Gordon Rattray, *The Doomsday Book*. Cleveland: World, 1970; Fawcett, 1971.

TAYLOR, J. A., *Climatic Resources and Economic Activity*. New York: Halsted Press, 1974.

TAYLOR, Theodore B., "Strategies for the Future," *Saturday Review/World*, December 14, 1974.

TAYLOR, Theodore B., and Charles C. HUMPSTONE, *The Restoration of the Earth*. New York: Harper & Row, 1973.

"Thaw in the Cold War," *Fortune*, April 1956.

"Thin Ice At The World's End," *Science News*, November 3, 1973.

THOMAS, Jack, "I'm Hungry," *East-West Journal*, May 1975.

THOMPSON, G. A., "A Plan for Converting the Sahara Desert Into a Sea," *Scientific American*, August 10, 1912.

THOMPSON, Louis M., "Weather Variability, Climatic Change, and Grain Production," *Science*, May 9, 1975.

THOMPSON, Philip D., Robert O'BRIEN et al., *Weather*. New York: Time-Life Books, 1965.

"Those Vaguely Sinister Skies," *Time*, August 25, 1975.

TINKER, Jon, "A Hydra-Headed Crisis," *New Scientist*, November 7, 1974; *Atlas World Press Review*, May 1975.

"To Feed the World: What to Do With Changing Climate," *Science*, November 9, 1973.

"To Move The Earth and Melt The Pole," *New York Times*, September 29, 1912; see editorial response, September 30; Riker rebuttal, October 3; and "Reader Suggests A World Fund to Carry Out Mr. Riker's Plan," October 4, 1912.

TORGERSON, Dial, "Chad Officials Confirm Libyan Seizure of Territory," *Los Angeles Times*, September 27, 1975.

TOTH, Robert C., "Russ Begin Spitsbergen Air Service," *Los Angeles Times*, September 10, 1975.

TOYNBEE, Arnold, "After the Age of Affluence . . . ," *London Observer*, April 14, 1974.

TRB (STROUT, Richard L.), "The Power to Decide Who Lives, Who Starves," *Los Angeles Times*, November 1, 1974.

TREWARTHA, Glenn T., *An Introduction to Climate*, 4th ed. New York: McGraw-Hill, 1968.

TROEBST, Cord-Christian, *Conquest of the Sea*. New York: Harper & Row, 1962.

"Trying to Match Up Glaciers," *Science News*, July 8, 1972.

TUOHY, William, "Is Bangladesh Past the Point of Saving?" *Los Angeles Times*, September 1, 1975.

TURNER, Barry E., "Interstellar Molecules," *Scientific American*, March 1973.

ULRICH, Roger K., "Solar Neutrinos and Variations in the Solar Luminosity," *Science*, November 14, 1975.

UNESCO, *Changes of Climate*. Proceedings of the Rome Symposium. Paris, 1963.

UNESCO Staff, "Sahel: Land of No Return?" *UNESCO Courier*, Special Issue, April 1975.

URBACH, Frederick, ed., *Biologic Effects of Ultraviolet Radiation*. Oxford: Pergamon Press, 1969.

U.S. Army War College, ed., *New Dynamics in National Strategy*. New York: Crowell, 1975.

U.S. Department of Commerce/NOAA, *Heat Wave*. Washington, D.C.: U.S. Government Printing Office, 1972.

U.S. House of Representatives, *Chemical-Biological Warfare: U.S. Policies and International Effects*. Hearings before the National Security Policy and Scientific Development Subcommittee of the Committee on Foreign Affairs, 91st Congress, First Session, November 18, 20; December 2, 9, 18, 19, 1969. Rep. Clement J. Zablocki, Wis., chrmn.

U.S. House of Representatives, *Prohibition of Weather Modification as a Weapon of War*. Hearings before the International Organizations Subcommittee of the Committee on International Relations, 94th Congress, First Session, July 29, 1975. Rep. Donald M. Frazer, Minn., chrm.

U.S. House of Representatives, *Weather Modification as a Weapon of War*. Hearings before the International Organizations and Movements Subcommittee of the Committee on Foreign Affairs, 93rd Congress, Second Session, September 24, 1974. Rep. Donald M. Fraser, Minn., chrmn.

"U.S.–Russ Bridge to Span Bering Straits Proposed," *Los Angeles Times*, March 25, 1960.

U.S. Senate, *Prohibiting Military Weather Modification*. Hearings before the Oceans and International Environment Subcommittee of the Committee on Foreign Relations, 92nd Congress, Second Session, July 26 and 27, 1972. Sen. Claiborne Pell, R.I., chrmn.

U.S. Senate, *Weather Modification*. Hearings before the Oceans and International Environment Subcommittee of the Committee on Foreign Relations, 93rd Congress, Second Session, January 25 and March 20, 1974. Sen. Claiborne Pell, R.I., chrmn.

"U.S., Soviets, Science: Joint Projects," *Science News*, November 2, 1974.

"U.S./U.S.S.R. Bering Sea Weather Study," *Science News*, December 9, 1972.

UTLAUT, William F., and Robert COHEN, "Modifying the Ionosphere with Intense Radio Waves," *Science*, October 15, 1971.

"Value Doubted on Bering Dam," *Los Angeles Times*, January 3, 1960.

VAN ALLEN, James A., "Interplanetary Particles and Fields," *Scientific American*, September 1975.

VAN VALEN, Leigh, "Climate and Evolutionary Rate," *Science*, December 26, 1969.

VAN WOERKOM, A. J. J., "The Astronomical Theory of Climate Changes," in Harlow SHAPLEY, ed., *Climatic Change*. Cambridge: Harvard University Press, 1953.

VEEH, N. Herbert, and John CHAPPELL, "Astronomical Theory of Climatic Change: Support from New Guinea," *Science*, February 6, 1970.

VETTER, Richard C., *Oceanography: The Last Frontier*. New York: Basic Books, 1973.

"Volcanoes and Ice Ages: A Link?" *Science News*, February 15, 1975.

VON HAGEN, Victor W., *Realm of the Incas*, rev. ed. New York: Mentor, 1961.

―――, *World of the Maya*. New York: Mentor, 1960.

VON HIPPLE, Frank, "The Nuclear Debate," *Bulletin of Atomic Scientists*, May 1975.

WADE, Nicholas, "Green Revolution (I): A Just Technology, Often Unjust in Use," *Science*, December 20, 1974.

―――, "Green Revolution (II)" *Science*, December 27, 1974.

―――, "International Agricultural Research," *Science*, May 9, 1975.

WAIT, James R., "Project Sanguine," *Science*, October 20, 1972.

WALSH, John, "U.S. Agribusiness and Agricultural Trends," *Science*, May 9, 1975.

WALTERS, Harry, "Difficult Issues Underlying Food Problems," *Science*, May 9, 1975.

WARD, Barbara, and René DUBOS, *Only One Earth*. New York: Norton, 1972.

WARREN, Charles, "Parson Malthus Tolls the Bell," *Sierra Club Bulletin*, March 1975.

WATKINS, N. D., and J. P. KENNETT, "Antarctic Bottom Water: Major Change in Velocity During the Late Cenozoic Between Australia and Antarctica," *Science*, August 27, 1971.

WATSON, Lyall, *Supernature*. New York: Anchor Press, 1973.

WEARE, Bryan C., et al., "Aerosol and Climate: Some Further Considerations," *Science*, November 29, 1974.

"Weather and Climate Modification: Progress and Problems," *Science*, August 17, 1973.

"Weather from Antarctic," *Science News Letter*, May 17, 1952.

"Weather Modification Research Under A Cloud," *Science*, September 27, 1974.

"The Weather Weapon: New Race With the Reds," *Newsweek*, January 13, 1958.

WEAVER, Paul H., "Making the U.N. Safe for Democracy," *Fortune*, November 1975.

WEICKMANN, Helmut K., "The Mitigation of Great Lakes Storms," in W. N. HESS, ed., *Weather and Climate Modification*. New York: Wiley, 1974.

WEINBERG, Janet H., "The Green Machine," *Science News*, April 5, 1975.

WEINSTEIN, Martin E., *Japan's Postwar Defense Policy, 1947–1968*. New York: Columbia University Press, 1971.

WEISBERG, Barry, *Ecocide In Indochina*. San Francisco: Canfield Press, 1970.

WENK, Edward, Jr., *The Politics of the Ocean*. Seattle: University of Washington Press, 1972.

WESTING, Arthur H., and E. W. PFEIFFER, "The Cratering of Indochina," *Scientific American*, May 1972.

WESTOFF, Charles F., "The Populations of the Developed Countries," *Scientific American*, September 1974.

WEXLER, Harry, "Modifying Weather on a Large Scale," *Science*, Ocotber 31, 1958.

——, "Radiation Balance of the Earth as a Factor in Climatic Change," in Harlow SHAPLEY, ed., *Climatic Change*. Cambridge: Harvard University Press, 1953.

——, "Volcanoes and World Climate," *Scientific American*, April 1952.

WHARTON, Clifton R., Jr., "The Green Revolution: Cornucopia or Pandora's Box?" *Foreign Affairs*, April 1969.

"Whatever Happened to Food Research?" *Science News*, February 15, 1975.

"Wheat Harvest," *New Republic*, September 30, 1972.

"When Is a Greenhouse Effect Not?" *Science News*, December 20 & 27, 1975.

WHITAKER, H. R., "Don't Bet on Weather Modification Yet," *Science Digest*, September 1972.

WHITESIDE, Thomas, *Defoliation*. New York: Ballantine, 1970.

——, *The Withering Rain*. New York: Dutton, 1971.

"Why Aerosols Are Under Attack," *Business Week*, February 17, 1975.

WIESNER, C. J., *Climate, Irrigation and Agriculture*. Mystic, Conn.: Verry Press, 1970.

WIESNET, Donald R., and Michael MATSON, *Monthly Winter Snowline Variation in the Northern Hemisphere from Satellite Records, 1966-75*. Washington, D.C.: National Oceanic and Atmospheric Administration, 1975.

WILCOX, Howard A., *Hothouse Earth*. New York: Praeger, 1975.

WILCOX, J. M., "Link Solar Wind and Weather," *Mechanical Engineering*, March 1974.

WILFORD, John Noble, "Scientists Are Critical of Rainmaking in War," *New York Times*, July 3, 1972.

WILKINSON, John, and Wendell MORDY, "The Coming Climate Change and Famine," *Center Report*, February 1975.

WILLIAMS, J., "Influence of Snowcover on the Atmospheric Circulation and Its Role in Climatic Change . . . ," *Journal of Applied Meteorology*, March 1975.

WILLIAMS, Jerome, et al., *Sea and Air*. Annapolis: United States Naval Institute, 1968.

WILLETT, Hurd C., "Atmospheric and Oceanic Circulation as Factors in Glacial-Interglacial Changes of Climate," in Harlow SHAPLEY, ed., *Climatic Change*. Cambridge: Harvard University Press, 1953.

——, "Cold Weather Ahead!" *Saturday Evening Post*, March 24, 1956.

WILLRICH, Mason, ed., *International Safeguards and Nuclear Industry*. Baltimore: Johns Hopkins Press, 1973.

——, "Terrorists Keep Out!" *Bulletin of Atomic Scientists*, May 1975.

WILLRICH, Mason, and Theodore B., TAYLOR, *Nuclear Theft: Risks and Safeguards*. Cambridge, Mass.: Ballinger, 1974.

"Will Widespread Drought Hit The Country Soon?" (Interview with Walter Orr Roberts), *U.S. News & World Report*, March 18, 1974.

WILSON, Mitchell, et al., *Energy*. New York: Life Science Library, 1963.

WILSON, Richard, "Our Best Energy Bet for the Future," *Harvard Magazine*, November 1973.

"Winds of the Past," *Science News*, January 1, 1972.

WINKLESS, Nels III, and Iben BROWNING, *Climate and the Affairs of Men*. New York: Harper's Magazine Press, 1975.

——, "Our Coming Climate," *Country Journal*, November 1975.

——, *Weather, Weapons and Wisdom*. New York: Harper's Magazine Press, 1976 (Tentative).

WITTWER, S. H., "Food Production: Technology and the Resource Base," *Science*, May 9, 1975.

WOFSY, Steven C., et al., "Freon Consumption: Implications for Atmospheric Ozone," *Science*, February 14, 1975.

WOLBACH, John, "The Insufficiency of Geographical Causes of Climatic Change," in Harlow SHAPLEY, ed. *Climatic Change*. Cambridge: Harvard University Press, 1953.

WOODBURY, David O., *The Great White Mantle*. New York: Viking, 1962.

WOODWARD, Kenneth L., et al., "What Comes Naturally," *Newsweek*, September 1, 1975.

"Worldwide Weather: Normal or Abnormal?" *Electrical World*, February 15, 1975.

WORTMAN, Sterling, "Agriculture in China," *Scientific American*, June 1975.

WRIGHT, H. E., Jr., "Natural Environment of Early Food Production North of Mesopotamia," *Science*, July 28, 1968.

WRIGLEY, G., *Tropical Agriculture*. London: Faber, 1969.

"Writing the Law on the Weather," *Business Week*, March 24, 1951.

YOUNG, Andrew and Louise, "Venus," *Scientific American*, September 1975.

YOUNG, Louise B., and H. Peyton YOUNG, "Pollution by Electrical Transmission," *Bulletin of Atomic Scientists*, December 1974; discussion, Harold N. SCHERER and L. B. YOUNG, September 1975.

ZELITCH, I., "Improving the Efficiency of Photosynthesis," *Science*, May 9, 1975.

Index